17th Topical Conference on Gas Utilization 2017

Topical Conference at the 2017 AIChE Spring Meeting and 13th Global Congress on Process Safety

San Antonio, Texas, USA
26 – 30 March 2017

ISBN: 978-1-5108-4130-7

Printed from e-media with permission by:

Curran Associates, Inc.
57 Morehouse Lane
Red Hook, NY 12571

CURRAN ASSOCIATES INC. proceedings.com

Some format issues inherent in the e-media version may also appear in this print version.

Printed by Curran Associates, Inc. (2017)

For permission requests, please contact AIChE
at the address below.

AIChE
120 Wall Street, FL 23
New York, NY 10005-4020

Phone: (800) 242-4363
Fax: (203) 775-5177

www.aiche.org

Additional copies of this publication are available from:

Curran Associates, Inc.
57 Morehouse Lane
Red Hook, NY 12571 USA
Phone: 845-758-0400
Fax: 845-758-2633
Email: curran@proceedings.com
Web: www.proceedings.com

TABLE OF CONTENTS

482030 AGILE Award Keynote Address

Abstract:

Factors in Selecting AGILE Honoree

1. Overall sustained excellent industry leadership in businesses that employ chemical engineers

2. Recognized for efforts in innovation and creating strategies that ensure success

3. Value of the keynote to the meeting attendees and the members of the Institute Schedule

Event	Time	Location
Continental Breakfast	7:00AM - 8:00AM	Henry Gonzalez Convention Center, Lila Cockrell Theatre
AGILE Keynote Address	8:00AM - 9:15AM	Henry Gonzalez Convention Center, Lila Cockrell Theatre

The 2017 AIChE Spring Meeting and 13th GCPS AGILE Keynote Address will be presented by AGILE award recipient John Y. Televantos, Partner, Arsenal Capital Partners.

John Televantos is a Partner at Arsenal Capital Partners, a middle market private equity firm focused on the specialty industrial, healthcare and specialty financial service sectors. Before joining Arsenal in 2006, Dr. Televantos was President of Aqualon Company, owned by Hercules, Inc.; President and Founder of Helios Chemical Company; and Chief Executive Officer for Foamex International.

Dr. Televantos began his career with Union Carbide in 1977 in silicones research. There he developed two key processes for the manufacture of Methyl Chloride and Siloxane polymers that were commercialized. These processes not only improved efficiency and reduced cost but also dramatically reduced waste. He moved through a variety of management positions there and later at ARCO Chemical, where he became vice president of strategy and development.

At ARCO John co-invented and led a team that developed and commercialized a breakthrough technology in polyol manufacture (a key raw material for polyurethanes) eliminating a whole process step along with waste and emissions, while at the same time opening the door to the creation of new higher-performance polyurethanes. The technology has since been licensed by most of the major global polyol manufacturers with significant waste reduction worldwide.

At Arsenal Capital, Dr. Televantos serves as co-head of the specialty industrial practice. During his tenure, he has been part of several acquisitions and successful exits in the chemicals and materials sectors. In addition to serving on the boards of several Arsenal-owned companies, he is lead director on the board of Balchem Corp.

482218 Energy-Industry Trends and Insights from the Wellhead to Downstream Products

Abstract:

HSB Solomon Associates LLC (Solomon) has conducted Comparative Performance Analyses™ (CPA™), also known as the Solomon studies, for over 35 years. These studies benchmark the Refining, Petrochemicals, Midstream, Upstream, and Electric Power industries, with the earliest studies starting in 1980. As part of these studies, Solomon gathers and analyzes data from around the globe on all phases of operations. Key Performance Areas include yields and losses, energy consumption, reliability and maintenance, personnel, efficiency, and financial performance. These benchmarking studies give Solomon clients valuable insights into the key performance areas of their facilities as well as the overall industry trends. The studies provide reliable information based on proprietary, patented, and time-tested methodologies, to foster the continuous improvement needed for companies to remain competitive.

The first Solomon studies began in 1980. Over the years, the studies have documented all of the ups and downs of the oil-and-gas-related industry as it reacted to changes in prices, markets, technology, and government regulations. For this paper, Solomon consultants have examined the trends and shared insights learned from their data-driven CPA on all energy-related industries from the wellhead to the downstream derivative chemical manufacturing, and all stages in-between.

479398 Depressurization of Process Equipment during Emergency or Planned Shutdown on a High Potential CO2-Solid-Forming Stream Utilizing the Latest Dynamic Simulation and Blowdown Models

Abstract:

Natural gas produced from many major reservoirs can contain significant amounts of carbon dioxide (CO2) and must be treated to meet typical specifications for pipelines or liquefaction plant feed. These treatment processes often require high-pressure operations forming highly concentrated CO2-rich streams. Pressure protection for these systems has been challenging to date because of the potential for CO2 solids generation upon pressure let down and the consequent potential for plugging.

Blowdown, the emergency or planned depressuring of process equipment, is a critical process safety operation. It may be necessary, in the event of a fire, leak, pipe rupture or other hazardous situation, as well as for a planned shutdown. Devices such as control valves, relief valves, restriction orifices, rupture disks, and safety valves transfer the potentially dangerous contents of process equipment to a safe lower-pressure location, or to the flare system for controlled combustion.

To ensure blowdown can be executed safety and effectively, a number of design concerns must be addressed, such as solid CO2 identification, low temperature (for both process and equipment material). Rapid depressuring and gas expansion can potentially putting equipment at risk of brittle fracture and if the construction material goes below its ductile-brittle transition temperature as well as potential plugging, due to CO2 solid formation. In addition, the entire pressure relief system, including safety valves, relief orifices, flare piping and knockout drums, must be sufficiently sized to handle the flowrates that occur during blowdown, in addition to the piping and capacity of the flare system.

For new installations, accurately predicting the minimum vessel wall temperature during blowdown is important for selecting the appropriate construction material, for helping reduce overdesign and consequently for lowering project cost. Similarly, having an accurate prediction of the maximum flow rate during blowdown reduces overdesign associated with the relief valve/network, without compromising on safety.

The paper will address the potential of solid CO2 formation based on commercially available software for blowdown and propose some mitigation plan with respect to solid CO2 formation.

Inherently Safer Piping – Best Practices

Steve Streblow
ACS Engineering
10914 Milano Ct.
Richmond, TX 77406
streblowsr@yahoo.com

Prepared for Presentation at
American Institute of Chemical Engineers
2017 Spring Meeting and 13th Global Congress on Process Safety
San Antonio, TX
March 26 – 29, 2017

Inherently Safer Piping – Best Practices

Steve Streblow
ACS Engineering
10914 Milano Ct.
Richmond, TX 77406
streblowsr@yahoo.com

Keywords: 49 CFR 192, 49 CFR 195, API 1102, ASME B31, buried, midstream, pipe, piping, pressure, relief, safety, upstream

Abstract

This paper presents lesser known items in the upstream and midstream industries important to onshore metal piping stress and overpressure protection.

Well known RAGAGEP presented are 49 CFR 195, 49 CFR 192 and the ASME piping codes B31.1/3/4/8. Specific items include corrosion allowance, equations, manufacturing tolerance, maximum allowable pressures, overpressure protection devices, stress values and thread depth. Inspection, testing and maintenance of over pressure protection devices is also discussed. The implications of B31.8 references to B31.1 for water and steam and B31.3 for air, lube oil and hydraulic power are discussed. Many times, design pressures are set below the code minimum design pressures, so those are presented for steam, feedwater, blow piping and drains.

Items detailed outside the CFRs and ASME piping code that should be taken into account, such as surface dead loads, surface live loads, and cyclic loads on buried pipe fatigue are also presented. Failure mechanisms, equations, and acceptance criteria for deflection, ovality, through wall bending, crushing, ring buckling, fatigue, longitudinal, thermal expansion and fluid transients are described. Sources of this information are API Recommended Practice 1102, Steel Pipelines Crossing Railroads; and Highways and American Lifelines Alliance, Guidelines for the Design of Buried Steel Pipe.

Using these presented methods results in inherently safer piping and safer pressure relief system design.

1 Introduction

In having worked with and for many upstream and midstream engineering and operating companies, I have seen that the onshore piping design (for both gases and liquids) is often not done in enough detail. I have always seen that the pipe wall thickness is checked for maximum internal operating pressure, but even then, there are some factors that are often times overlooked.

For gas piping, the category (production, gathering, transmission or distribution) is first determined, and then a determination if the piping is regulated or not is made. The main regulations for gas piping, 49 CFR 192, the minimum requirement, is discussed.

The main regulation for liquid piping, 49 CFR 195, the minimum requirement, is discussed.

Then, some best practices are presented, starting with ASME B31.8 for Gas Piping and Stations such as Compression. The best practice for liquid piping, ASME B31.4 is presented. The sections of ASME B31.1 and B31.3 which are best practices for Compressor station piping are presented.

Some best practices for piping, regardless of fluid and location are presented, starting with API 1102 and then finishing with Guidelines for the Design of Buried Steel Pipe.

Using these documents allows for many more factors such as corrosion allowance, manufacturing tolerance, stress values and thread depth to be taken into account.

Surface dead loads, surface live loads, and cyclic loads on buried pipe fatigue are also presented. Failure mechanisms, equations, and acceptance criteria for deflection, ovality, through wall bending, crushing, ring buckling, fatigue, longitudinal, thermal expansion and fluid transients are described.

Using these presented methods results in inherently safer piping and safer pressure relief system design.

2 What is the Gas Piping Category?

For natural gas piping, one of the first steps is to determine if the pipeline is production, gathering, transmission or distribution. Reference 1, section 2.2, is used to determine the split between production and gathering.

The gathering classification can't extend beyond the first natural gas processing plant, when it switches to transmission. Transmission then eventually switches to distribution.

3 Regulated Gas Gathering or Not?

In reference 2, there is a parameter called the class location, which accounts for risk due to buildings and occupancy near the pipeline. The class location ranges from 1, for lowest risk to 4 for highest risk. Details in reference 2, paragraph 192.5.

If the gas piping is gathering, one must then determine if it is regulated, per reference 2, paragraph 192.8. Regulated gathering depends on the hoop stress, class location and safety buffer. Regulated gathering lines have to follow certain aspects of reference 2.

4 CFR 192 for Gas Piping – Part 1

So, one needs to follow reference 2 for regulated gathering and transmission piping.

Per reference 2, paragraph "192.103 General. Pipe must be designed with sufficient wall thickness, or must be installed with adequate protection, to withstand anticipated external pressures and loads that will be imposed on the pipe after installation." However, reference 2 does not provide any formulas or methods to satisfy this requirement. This is one area that I have commonly seen overlooked.

Per reference 2, paragraph "192.105 Design formula for steel pipe. (a) The design pressure for steel pipe is determined in accordance with the following formula:"

$P = (2\ S \times t/D) \times F \times E \times T$, where

P =Design pressure in pounds per square inch gage.
S =Yield strength in pounds per square inch determined in accordance with §192.107.
t = Nominal wall thickness of the pipe in inches. Additional wall thickness required for concurrent external loads in accordance with §192.103 may not be included in computing design pressure.
D =Nominal outside diameter of the pipe in inches.
F =Design factor determined in accordance with §192.111.
E =Longitudinal joint factor determined in accordance with §192.113.
T =Temperature derating factor determined in accordance with §192.115.

Per reference 2, paragraph "192.107 Yield strength (S) for steel pipe."…"The yield strength to be used in the design formula in §192.105 is the SMYS stated in the listed specification."

Per reference 2, paragraph "192.111 Design factor (F) for steel pipe.

Table 1. Design factor (F) for steel pipe.

Class location	Design factor (F)
1	0.72
2	0.60
3	0.50
4	0.40

A design factor of 0.60 or less must be used in the design formula in paragraph 192.105 for steel pipe in Class 1 locations that: (1) Crosses the right-of-way of an unimproved public road, without a casing: (2) Crosses without a casing, or makes a parallel encroachment on, the right-of-way of either a hard surfaced road, a highway, a public street, or a railroad: (3) Is supported by a vehicular, pedestrian, railroad, or pipeline bridge: or (4) Is used in a fabricated assembly, (including separators, mainline valve assemblies, cross-connections, and river crossing headers) or is used within five pipe diameters in any direction from the last fitting of a fabricated assembly, other than a transition piece or an elbow used in place of a pipe bend which is not associated with a fabricated assembly. (c) For Class 2 locations, a design factor of 0.50, or less, must be used in the design formula in §192.105 for uncased steel pipe that crosses the right-of-way of a hard surfaced road, a highway, a public street, or a railroad. (d) For Class 1 and Class 2 locations, a de-sign factor of 0.50, or less, must be used in the design formula in §192.105 for– (1) Steel pipe in a compressor station, regulating station, or measuring station." These more specific design factors for specific locations and types of facilities is one area that I have commonly seen overlooked.

Per reference 2, "§192.113 Longitudinal joint factor (E) for steel pipe. The longitudinal joint factor to be used in the design formula in §192.105 is determined in accordance with the following table:"

5 E factor comparison

The E factor table from 49 CFR 192 was combined with the E factor table from 49 CFR 195 and the E factor from ASME B31.8 and is presented immediately below. Comparison of the three tables shows that there are a few differences.

Table 2. Factor (E) for steel pipe.

<table>
<tr><th>Specification</th><th>Pipe Class</th><th>49 CFR 192 Longitudinal Joint Factor (E)</th><th>49 CFR 195 Seam Joint Factor (E)</th><th>ASME B31.8</th></tr>
<tr><td rowspan="4">ASTM A53/A53M</td><td>Seamless</td><td colspan="3">1.00</td></tr>
<tr><td>Electric resistance welded</td><td colspan="3">1.00</td></tr>
<tr><td>Furnace lap welded</td><td>NA</td><td>0.80</td><td>NA</td></tr>
<tr><td>Furnace butt welded</td><td colspan="3">0.60</td></tr>
<tr><td>ASTM A106</td><td>Seamless</td><td colspan="3">1.00</td></tr>
<tr><td>ASTM A134</td><td>Electric-fusion arc-welded</td><td colspan="2">NA</td><td>0.80</td></tr>
<tr><td>ASTM A135</td><td>Electric resistance-welded</td><td colspan="2">NA</td><td>1.00</td></tr>
<tr><td>ASTMA139</td><td>Electric-fusion arc-welded</td><td colspan="2">NA</td><td>0.80</td></tr>
<tr><td rowspan="3">ASTM A333/A333M</td><td>Seamless</td><td>NA</td><td>1.00</td><td>1.00</td></tr>
<tr><td>Electric resistance welded</td><td>1.00</td><td>NA</td><td>1.00</td></tr>
<tr><td>Welded</td><td>NA</td><td>1.00</td><td>NA</td></tr>
<tr><td>ASTM A381</td><td>Double submerged arc welded</td><td colspan="3">1.00</td></tr>
<tr><td rowspan="2">ASTM A671</td><td>Electric-fusion welded
Classes 13, 23, 33, 43, 53</td><td colspan="2">1.00</td><td>0.80</td></tr>
<tr><td>Electric-fusion welded
Classes 12, 22, 32, 42, 52</td><td colspan="3">1.00</td></tr>
<tr><td rowspan="2">ASTM A672</td><td>Electric-fusion welded
Classes 13, 23, 33, 43, 53</td><td colspan="2">1.00</td><td>0.80</td></tr>
<tr><td>Electric-fusion welded
Classes 12, 22, 32, 42, 52</td><td colspan="3">1.00</td></tr>
</table>

Table 2. Factor (E) for steel pipe. (continued)

Specification	Pipe Class	49 CFR 192 Longitudinal Joint Factor (E)	49 CFR 195 Seam Joint Factor (E)	ASME B31.8
ASTM A691	Electric-fusion welded Classes 13, 23, 33, 43, 53	1.00		0.80
	Electric-fusion welded Classes 12, 22, 32, 42, 52	1.00		
ASTM A984	Electric-resistance-welded	NA		1.00
ASTM A1005	Double submerged-arc-welded	NA		1.00
ASTM A1006	Laser beam welded	NA		1.00
API 5L	Seamless	1.00		
	Electric resistance welded	1.00		
	Electric flash welded	1.00		
	Submerged arc welded	1.00		
	Furnace lap welded	NA	0.80	NA
	Furnace butt welded	0.60		
Other	Pipe over 4 inches (102 millimeters)	0.80	NA	
Other	Pipe 4 inches (102 millimeters) or less	0.60	NA	

6 CFR 192 for Gas Piping – Part 2

Per reference 2, "192.115 Temperature derating factor (T) for steel pipe. The temperature derating factor to be used in the design formula in §192.105 is determined as follows:

Table 3. Temperature derating factor (T) for steel pipe.

Gas Temperature in degrees Fahrenheit	Temperature derating factor (T)
250 or less	1.000
300	0.967
350	0.933
400	0.900
450	0.867

For intermediate gas temperatures, the derating factor is determined by interpolation."

Per reference 2, "§192.149 Standard fittings. (a) The minimum metal thickness of threaded fittings may not be less than specified for the pressures and temperatures in the applicable standards referenced in this part, or their equivalent." However, reference 2 does not provide any formulas or methods to satisfy this requirement. This is one area that I have commonly seen overlooked.

Per reference 2, "§192.159 Flexibility. Each pipeline must be designed with enough flexibility to prevent thermal expansion or contraction from causing excessive stresses in the pipe or components, excessive bending or unusual loads at joints, or undesirable forces or moments at points of connection to equipment, or at anchorage or guide points." However, reference 2 does not provide any formulas or methods to satisfy this requirement. This is one area that I have commonly seen overlooked.

Per reference 2, "§192.161 Supports and anchors.

(a) Each pipeline and its associated equipment must have enough anchors or supports to: (1) Prevent undue strain on connected equipment; (2) Resist longitudinal forces caused by a bend or offset in the pipe; and, (3) Prevent or damp out excessive vibration.

(b) Each exposed pipeline must have enough supports or anchors to protect the exposed pipe joints from the maximum end force caused by internal pressure and any additional forces caused by temperature expansion or contraction or by the weight of the pipe and its contents.

(c) Each support or anchor on an exposed pipeline must be made of durable, noncombustible material and must be designed and installed as follows: (1) Free expansion and contraction of the pipeline between supports or anchors may not be restricted. (2) Provision must be made for the service conditions involved." However, reference 2 does not provide any formulas or methods to satisfy this requirement. This is one area that I have commonly seen overlooked.

It is important to note that reference 2 describes minimum standards, not best practices. Read on to learn about best practices.

7 CFR 195 for Liquid Piping – Part 1

Hazardous liquids transportation pipelines are often required to follow reference 3 for several aspects, including design.

From reference 3, "§195.1 Which pipelines are covered by this part?" "(2) Transportation through any pipeline, other than a gathering line, that has a maximum operating pressure (MOP) greater than 20-percent of the specified minimum yield strength;" "(4) Transportation of petroleum in any of the following onshore gathering lines: (i) A pipeline located in a non-rural area; (ii) To the extent provided in §195.11, a regulated rural gathering line defined in §195.11."

From reference 3, "(b) Excepted. This part does not apply to any of the following:" "(8) Transportation of a hazardous liquid or carbon dioxide through onshore production (including flow lines), refining, or manufacturing facilities or storage or inplant piping systems associated with such facilities;"

From reference 3, "§195.11 What is a regulated rural gathering line and what requirements apply?...(a) Definition. As used in this section, a regulated rural gathering line means an onshore gathering line in a rural area that meets all of the following criteria—(1) Has a nominal diameter from 6⅝ inches (168 mm) to 8⅝ inches (219.1 mm);...(3) Operates at a maximum pressure established under §195.406 corresponding to—(i) A stress level greater than 20-percent of the specified minimum yield strength of the line pipe"

From reference 3, "§195.106 Internal design pressure. (a) Internal design pressure for the pipe in a pipeline is determined in accordance with the following formula:
P = (2 S x t/D) x E x F", where

P = Internal design pressure in p.s.i. gage.
S = Yield strength in pounds per square inch determined in accordance with paragraph (b) of this section.
t = Nominal wall thickness of the pipe in inches.
D = Nominal outside diameter of the pipe in inches.
E = Seam joint factor determined in accordance with paragraph (e) of this section.
F = A design factor of 0.72

8 Comparing Wall Thickness for Gas and Liquid Piping

Comparing the internal design pressure formulas of 49 CFR 192 and 195 shows that the formula are slightly different. 49 CFR 192 design factors range from 0.72 to 0.4, while 49 CFR 195 design factor is set at 0.72. Therefore it can be seen that for a given internal design pressure, with the same piping, the gas piping oftentimes needs to have a thicker wall. 49 CFR 192 temperature derating factor ranges from 1.000 to 0.867. while 49 CFR 195 does not have a temperature derating factor, therefore, for a given operating temperature with the same piping, the gas piping needs to have a thicker wall.

9 CFR 195 for Liquid Piping – Part 2

From reference 3, "(b) The yield strength to be used in determining the internal design pressure under paragraph (a) of this section is the specified minimum yield strength."

From reference 3, "(d) The minimum wall thickness of the pipe may not be less than 87.5 percent of the value used for nominal wall thickness in determining the internal design pressure under paragraph (a) of this section. In addition, the anticipated external loads and external pressures that are concurrent with internal pressure must be considered in accordance with §§195.108 and 195.110 and, after determining the internal design pressure, the nominal wall thickness must be increased as necessary to compensate for these concurrent loads and pressures." However, reference 3 does not provide any formulas or methods to satisfy this requirement. This is one area that I have commonly seen overlooked.

From reference 3, "(e) The seam joint factor used in paragraph (a) of this section is determined in accordance with the following table:" See Table 2 above.

Per reference 3, "§195.108 External pressure. Any external pressure that will be exerted on the pipe must be provided for in designing a pipeline system." However, reference 3 does not provide any formulas or methods to satisfy this requirement. This is one area that I have commonly seen overlooked.

Per reference 3, "§195.110 External loads. (a) Anticipated external loads (e.g.), earthquakes, vibration, thermal expansion, and contraction must be provided for in designing a pipeline system. In providing for expansion and flexibility, §419 of ASME/ANSI B31.4 must be followed. (b) The pipe and other components must be supported in such a way that the support does not cause excess localized stresses. In designing attachments to pipe, the added stress to the wall of the pipe must be computed and compensated for." However, reference 3 does not provide any formulas or methods to satisfy this requirement. I This is one area that I have commonly seen overlooked.

It is important to note that reference 3 describes minimum standards, not best practices. Read on to learn about best practices.

10 Best Practice For Gas Piping and Stations such as Compression – ASME B31.8

What if the gas pipeline or facility, such as production or non-regulated gathering, is not subject to reference 2? How does one account for factors and loads not detailed in CFR 192? One can then follow reference 4, which has the following applicability "piping transporting products that are predominately gas between sources and terminals, including compressor, regulating, and metering stations; gas gathering pipelines."

From reference 4, paragraph "802.1 Scope…(b) This Code does not apply to…(4) piping in oil refineries or natural gasoline extraction plants, gas treating plant piping other than the main gas stream piping in dehydration, and all other processing plants installed as part of a gas transmission system, gas manufacturing plants, industrial plants, or mines (See other applicable sections of the ASME Code for Pressure Piping, B31.) (5) vent piping to operate at substantially atmospheric pressures for waste gases of any kind."

From reference 4, paragraph "833.2 Calculation of Longitudinal Stress Components
(a) The longitudinal stress due to internal pressure in restrained pipelines is
$S_P = 0.3S_H$, where S_H is the hoop stress, psi
(b) The longitudinal stress due to internal pressure in unrestrained pipeline is
$S_P = 0.5S_H$, where S_H is the hoop stress, psi
(c) The longitudinal stress due to thermal expansion in restrained pipe is $S_T = E\ a\ (T_1 - T_2)$, where
E = the elastic modulus, psi, at the ambient temperature, which can be found in reference 4, Table 832.5-1.
T_1 = the pipe temperature at the time of installation, tie-in, or burial, °F
T_2 = the warmest or coldest pipe operating temperature, °F
a = the coefficient of thermal expansion, 1/°F, which can be found in reference 4, Table 832.1-1
If a section of pipe can operate either warmer or colder than the installed temperature, both conditions for T_2 may need to be examined." This is one area that I have commonly seen overlooked.

From reference 4, paragraph "833.3 Summation of Longitudinal Stress in Restrained Pipe
(a) The net longitudinal stresses in restrained pipe are $S_L = S_P + S_T + S_X + S_B$
Note that S_B, S_L, S_T, or S_X can have negative values.
(b) The maximum permitted value of $|S_L|$ is 0.9ST, where S is the specified minimum yield strength, psi, per para. 841.1.1(a), and T is the temperature derating factor per para. 841.1.8." This is one area that I have commonly seen overlooked.

From reference 4, paragraph "833.4 Combined Stress for Restrained Pipe
(a) The combined biaxial stress state of the pipeline in the operating mode is evaluated using the calculation in either (1) or (2) below. (1) max. $\{|S_H - S_L|, |S_H|, |S_L|\}$ or (2) $[S_L^2 - S_LS_H + S_H^2]^{1/2}$
The maximum permitted value for the combined biaxial stress is kST where S is the specified minimum yield strength, psi, per para. 841.1.1(a), T is the temperature derating factor per para. 841.1.8, and k is defined in (b) and (c) below. (b) For loads of long duration, the value of k shall not exceed 0.90. (c) For occasional nonperiodic loads of short duration, the value of k shall not exceed 1.0. (d) S_L in (a) above is calculated considering both the tensile and compressive values of S_B. (e) Stresses induced by loads that do not occur simultaneously need not be considered to be additive. (f) The biaxial stress evaluation described above applies only to straight sections of pipe." This is one area that I have commonly seen overlooked.

From reference 4, paragraph "833.6 Summation of Longitudinal Stresses in Unrestrained Pipe
(a) The net longitudinal stress in unrestrained pipe is $S_L = S_P + S_X + S_B$, psi (b) The maximum permitted longitudinal stress in unrestrained pipe is $S_L \leq 0.75ST$, where S is the specified minimum yield strength, psi, per para. 841.1.1(a), and T is the temperature derating factor per para. 841.1.8." This is one area that I have commonly seen overlooked.

From reference 4, paragraph "833.8 Flexibility Stresses and Stresses Due to Periodic or Cyclic Fatigue Loading (b) The cyclic stress range $S_E <= S_A$, where
$S_A = f [1.25 (S_c + S_h) - S_L]$, where
$f = 6N^{-0.2} \leq 1.0$
N = equivalent number of cycles during the expected service life of the piping system
$S_c = 0.33\ S_uT$ at the minimum installed or operating temperature
$S_h = 0.33\ S_uT$ at the maximum installed or operating temperature
S_L = longitudinal stress calculated according to para. 833.6(a), psi
S_u = specified minimum ultimate tensile strength, psi
T = temperature derating factor per para. 841.1.8"
This is one area that I have commonly seen overlooked.

Per reference 4, paragraph "833.9 Local Stresses (b) The maximum allowable sum of circumferential stress due to internal pressure and circumferential through-wall bending stress caused by surface vehicle loads or other local loads is 0.9ST, where S is the specified minimum yield strength, psi, per para. 841.1.1(a), and T is the temperature derating factor per para. 841.1.8." This is one area that I have commonly seen overlooked.
Per reference 4, paragraph "833.10 Buckling and Lateral Instability. In order to prevent buckling in the form of wrinkling of the pipe wall or lateral instability, the maximum allowable net compressive stress is 2⁄3 of the critical buckling stress estimated using a suitable stability criterion." This is one area that I have commonly seen overlooked.

Per reference 4, see "Table 841.1.6-2 Design Factors for Steel Pipe Construction". Consider these in addition to those from reference 2.

From reference 4, paragraph "841.1.9 Additional Design Information or Instructions…
(g) Pressure/Flow Control Facilities…
(1) Overpressure protection shall be provided by the use of (-a) a monitor regulator in series with a controlling regulator (each regulator run). (-b) adequately sized relief valve(s) downstream of the controlling regulator(s). (-c) overpressure shutoff valve(s) upstream or downstream of the controlling regulator(s). Installation of alarm devices that indicate primary (controlling) regulator failure is useful and should be considered for monitoring regulator systems…
(k) Other (Nonmandatory) Considerations for Pressure Cycle Fatigue of Longitudinal Pipe Seams. Line pipe longitudinal seams are not generally regarded as susceptible to fatigue due to operational pressure cycles in most natural gas service. Evaluation of the potential for fatigue crack growth due to pressure cycles is recommended for pipe containing longitudinal seams where the expected lifetime accumulation of full MAOP cycles may exceed the following number of occurrences: $N = 6.0x10^{17} / ((F\ x\ S)^3\ x\ t^{0.5})$ where
F = design factor from Table 841.1.6-1
N = equivalent number of cycles during the expected service life of the piping system
S = specified minimum yield strength, psi
t = nominal wall thickness, in."
This is one area that I have commonly seen overlooked.

From reference 4, paragraph "843.3.4 Pressure-Limiting Requirements in Compressor Stations

(a) Pressure Relief. Pressure relief or other suitable protective devices of sufficient capacity and sensitivity shall be installed and maintained to ensure that the maximum allowable operating pressure of the station piping and equipment is not exceeded by more than 10%.
(b) Pressure Relief Valves. A pressure relief valve or pressure-limiting device, such as a pressure switch or unloading device, shall be installed in the discharge line of each positive displacement transmission compressor between the gas compressor and the first discharge block valve. If a pressure relief valve is the primary overprotection device, then the relieving capacity shall be equal to or greater than the capacity of the compressor.
(c) Venting. Vent lines provided to exhaust the gas from the pressure relief valves to atmosphere shall be extended to a location where the gas may be discharged without undue hazard. Vent lines shall have sufficient capacity so that they will not inhibit the performance of the relief valve. For additional design considerations, see para. 841.1.9(d)." This is one area that I have commonly seen overlooked.

From reference 4, paragraph "843.4.3 Air Piping System. (a) All air piping within gas compressing stations shall be constructed in accordance with ASME B31.3". This is one area that I have commonly seen overlooked.

From reference 4, paragraph "843.4.4 Lubricating Oil Piping. All lubricating oil piping within gas compressing stations shall be constructed in accordance with ASME B31.3." This is one area that I have commonly seen overlooked.

From reference 4, paragraph "843.4.5 Water Piping. All water piping within gas compressing stations shall be constructed in accordance with ASME B31.1." This is one area that I have commonly seen overlooked.

From reference 4, paragraph "843.4.6 Steam Piping. All steam piping within gas compressing stations shall be constructed in accordance with ASME B31.1." This is one area that I have commonly seen overlooked.

From reference 4, paragraph "843.4.7 Hydraulic Piping. All hydraulic power piping within gas compressing stations shall be constructed in accordance with ASME B31.3." This is one area that I have commonly seen overlooked.

From reference 4, paragraph "844.5 General Provisions Applicable to Both Pipe-Type and Bottle-Type Holders...Relief valves shall be installed in accordance with provisions of this Code that will have relieving capacity adequate to limit the pressure imposed on the filling line and thereby on the storage holder to 100% of the design pressure of the holder or to the pressure that produces a hoop stress of 75% of the specified minimum yield strength of the steel, whichever is the lesser." This is one area that I have commonly seen overlooked.

Paragraph "845.3 Requirements for Design of Pressure Relief and Pressure-Limiting Installations. (c) The discharge stacks, vents, or outlet ports of all pressure relief devices shall be located where gas can be discharged into the atmosphere without undue hazard." This is one area that I have commonly seen overlooked.

From reference 4, paragraph "845.4.1 Required Capacity of Pressure-Relieving and Pressure-Limiting Stations...(a) Each pressure relief station, pressure-limiting station, or group of such stations installed to protect a piping system or pressure vessel shall have sufficient capacity and shall be set to operate to prevent the pressure from exceeding the following levels: (1) Systems With Pipe or Pipeline Components Operating at Hoop Stress Levels Over 72% of the SMYS. The required capacity is the maximum allowable operating pressure plus 4%. (2) Systems With Pipe or Pipeline Components Operating at Hoop Stress Levels at or Below 72% of the SMYS Other Than in Low-Pressure Distribution Systems. The required capacity is the lesser of the following two items: (-a) the maximum allowable operating pressure plus 10% (-b) the pressure that produces a hoop stress of 75% of the specified minimum yield strength." This is one area that I have commonly seen overlooked.

From reference 4, paragraph "845.4.2 Proof of Adequate Capacity and Satisfactory Performance of Pressure-Limiting and Pressure Relief Devices. Where the safety device consists of an additional regulator that is associated with or functions in combination with one or more regulators in a series arrangement to control or limit the pressure in a piping system, suitable checks shall be made. These checks shall be conducted to determine that the equipment will operate in a satisfactory manner to prevent any pressure in excess of the established maximum allowable operating pressure of the system, should any one of the associated regulators malfunction or remain in the wide open position." This is one area that I have commonly seen overlooked.

From reference 4, paragraph "849.2 Steel Service Lines. (b) Steel service pipe shall be designed in accordance with the requirements of paras. 841.1.1 and 841.1.9(a). Where the pressure is less than 100 psig, the steel service pipe shall be designed for at least 100 psig pressure." This is one area that I have commonly seen overlooked.

From reference 4, paragraph "853.1 Compressor Station Maintenance...853.1.2 Inspection and Testing of Relief Valves. All pressure-relieving devices in compressor stations shall be inspected and/or tested in accordance with para. 853.3, and all devices except rupture disks shall be operated periodically to determine that they open at the correct set pressure. Any defective or inadequate equipment found shall be promptly repaired or replaced. All remote control shutdown devices shall be inspected and tested at least annually to determine that they function properly." This is one area that I have commonly seen overlooked.

From reference 4, paragraph "853.3 Maintenance of Pressure-Limiting and Pressure-Regulating Stations...853.3.1 Condition and Adequacy. All pressure limiting stations, relief devices, and other pressure regulating stations and equipment shall be subject to systematic, periodic inspections and suitable tests, or reviewed to determine that they are...(2) At least once each calendar year, a review shall be made to ensure that the combined capacity of the relief devices on a piping system or facility is adequate to limit the pressure at all times to values prescribed by this Code. This review should be based on the operating conditions that create the maximum probable requirement for relief capacity in each case, even though such operating conditions actually occur infrequently and/or for only short periods of time. If it is determined that the relieving equipment is of insufficient capacity, steps shall be taken to install new or additional equipment to provide adequate capacity." This is one area that I have commonly seen overlooked.

From reference 4, paragraph "853.3.3 Stop Valves (a) An inspection and/or test of stop valves shall be made to determine that the valves will operate and are correctly positioned. (Caution shall be used to avoid any undesirable effect on pressure during operational checks.) The following shall be included in the inspection and/or test: (1) station inlet, outlet, and bypass valves (2) relief device isolating valves (3) control, sensing, and supply line valves (b) The final inspection procedure shall include the following: (1) a check for proper position of all valves. Special attention shall be given to regulator station bypass valves, relief device isolating valves, and valves in control, sensing, and supply lines. (2) restoration of all locking and security devices to proper position." This is one area that I have commonly seen overlooked.

From Reference 4, see "MANDATORY APPENDIX D, SPECIFIED MINIMUM YIELD STRENGTH FOR STEEL PIPE, COMMONLY USED IN PIPING SYSTEMS"

11 Best Practice for Liquid Piping – ASME B31.4

What if the liquid pipeline or facility, such as production or non-regulated gathering, is not subject to reference 3? How does one account for factors and loads not detailed in CFR 195? One can then follow reference 5, which has the following applicability "piping transporting products that are predominately liquid between plants and terminals and within terminals, pumping, regulating, and metering stations."

Reference 5 shows that various categories of loads need to be considered, such as paragraph "401.1.2 Sustained Loads…401.1.3 Occasional Loads…401.1.4 Construction Loads and 401.1.5 Transient Loads." However, reference 5 does not provide any formulas or methods to satisfy this requirement. This is one area that I have commonly seen overlooked.

Reference 5 shows some subcategories of loads needing consideration, such as paragraph "401.2.2.2 Internal Design Pressure…401.2.2.3 External Hydrostatic Pressure…401.2.2.4 Weight Effects…401.2.2.5 Residual Loads…401.2.2.6 Subsidence…401.2.3.1 Earthquakes, 401.2.3.2 Wind Loads…401.2.3.3 Ice Loads…401.2.3.4 Road and Rail Traffic…401.2.3.5 Vibration…401.2.3.6 Waves and Currents…401.2.3.7 Temperature Effects…401.2.4.1 Installation Loads and 401.2.4.2 Hydrostatic Testing." However, reference 5 does not provide any formulas or methods to satisfy this requirement. This is one area that I have commonly seen overlooked.

From reference 5, paragraph "401.3 Combining of Loads" needs to be considered. This is one area that I have commonly seen overlooked.

From reference 5, paragraph "402 CALCULATION OF STRESSES…402.1 General…Circumferential, longitudinal, shear, and equivalent stresses shall be considered, taking into account stresses from all relevant sustained, occasional, construction, and transient loads, including vibration, resonance, and subsidence." This is one area that I have commonly seen overlooked.

From reference 5, paragraph "403.2 Criteria for Pipe Wall Thickness and Allowances…403.2.1 Criteria. The nominal wall thickness of straight sections of steel pipe shall be equal to or greater than t_n determined in accordance with the following equation: $t_n \geq t + A$, where

A = sum of allowances for threading, grooving, corrosion, and erosion as required in paras. 403.2.2 through 403.2.4, and increase in wall thickness if used as protective measure in para. 403.1

t_n = nominal wall thickness satisfying requirements for pressure and allowances

t = pressure design wall thickness"

This is one area that I have commonly seen overlooked.

In reference 5, see "Table 403.3.1-1 Allowable Values for Pipeline System Stresses." This is one area that I have commonly seen overlooked.

From reference 5, Paragraph "452.2 Controls and Protective Equipment (a) Controls and protective equipment, including pressure-limiting devices, regulators, controllers, relief valves, and other safety devices, shall be subjected to systematic periodic inspections and tests, at least annually, except as provided in para. 452.2(b), to determine that they are (1) in good mechanical condition (2) adequate from the standpoint of capacity and reliability of operation for the service in which they are employed (3) set to function at the correct pressure (4) properly installed and protected from foreign materials or other conditions that might prevent proper operation (b) Relief valves on pressure storage vessels containing LPG or liquid anhydrous ammonia shall be subjected to tests at least every 5 yr." This is one area that I have commonly seen overlooked.

12 Best Practice for Compressor Station Piping- ASME B31.1

From reference 4, paragraph "843.4.5 Water Piping. All water piping within gas compressing stations shall be constructed in accordance with ASME B31.1." This is one area that I have commonly seen overlooked.

From reference 4, paragraph "843.4.6 Steam Piping. All steam piping within gas compressing stations shall be constructed in accordance with ASME B31.1." This is one area that I have commonly seen overlooked.

So, to follow the best practice for compressor station water liquid and vapor piping, refer to reference 7, which has the following applicability "piping typically found in electric power generating stations, in industrial and institutional plants, geothermal heating systems, and central and district heating and cooling systems."

From reference 7, paragraph, "102.4.2 Threading and Grooving. The calculated minimum thickness of piping (or tubing) which is to be threaded shall be increased by an allowance equal to thread depth; dimension h of ASME B1.20.1 or equivalent shall apply. For machined surfaces or grooves, where the tolerance is not specified, the tolerance shall be assumed to be 1⁄64 in. (0.40 mm) in addition to the specified depth of cut. The requirements of para. 104.1.2(C) shall also apply." This is one area that I have commonly seen overlooked.

From reference 7, see "Table 102.4.3 Weld Joint Efficiency Factors"

From reference 7, paragraph "104.1.2 Straight Pipe Under Internal Pressure, Design pressure shall not exceed $P = 2SE(t_m - A) /(D_o - 2y(t_m - A))$, where
P = internal design pressure, psig
SE = maximum allowable stress in material due to internal pressure and joint efficiency at the design temperature, psi. The value of SE shall not exceed that given in Appendix A, for the respective material and design temperature. These values include the weld joint efficiency E.
tm = minimum required wall thickness, in. If pipe is ordered by its nominal wall thickness, the manufacturing tolerance on wall thickness must be taken into account. After the minimum pipe wall thickness tm is determined, this minimum thickness shall be increased by an amount sufficient to provide the manufacturing tolerance allowed in the applicable pipe specification or required by the process. The next heavier commercial wall thickness shall then be selected from thickness schedules such as contained in ASME B36.10M or from manufacturers' schedules for other than standard thickness. (A.1.4) Where ends are subject to forming or machining for jointing, the wall thickness of the pipe, tube, or component after such forming or machining shall not be less than tm minus the amount provided for removal by para. 104.1.2 (A.6.1).
A = additional thickness, in. to compensate for material removed in threading, grooving, etc., required to make a mechanical joint.
Do = outside diameter of pipe, in. For design calculations, the outside diameter of pipe as given in tables of standards and specifications shall be used in obtaining the value of tm. When calculating the allowable working pressure of pipe on hand or in stock, the actual measured outside diameter and actual measured minimum wall thickness at the thinner end of the pipe may be used to calculate this pressure.
y = coefficient having values as given in Table 104.1.2(A)". This is a temperature derating factor.

From reference 7, "(C.1) Where steel pipe is threaded and used for steam service at pressure above 250 psi or for water service above 100 psi with water temperature above 220°F, the pipe shall be seamless having the minimum ultimate tensile strength of 48,000 psi and a weight at least equal to Schedule 80 of ASME B36.10M." This is one area that I have commonly seen overlooked.

From reference 7, paragraph "104.1.3 Straight Pipe Under External Pressure. For determining wall thickness and stiffening requirements for straight pipe under external pressure, the procedures outlined in UG-28, UG-29, and UG-30 of Section VIII, Division 1 of the ASME Boiler and Pressure Vessel Code shall be followed." This is one area that I have commonly seen overlooked.

From reference 7, see paragraph "104.8.1 Stress Due to Sustained Loads formula." This is one area that I have commonly seen overlooked.

From reference 7, see paragraph "104.8.2 Stress Due to Occasional Loads formula." This is one area that I have commonly seen overlooked.

From reference 7, see paragraph "104.8.3 Thermal Expansion Stress Range." This is one area that I have commonly seen overlooked.

From reference 7, paragraph "122.1.2 Steam Piping. The design pressure shall not be taken at less than 100 psig for any condition of service or material." This is one area that I have commonly seen overlooked.

From reference 7, paragraph "122.1.3 Feedwater Piping. For piping from the boiler to and including the required stop valve and the check valve, the minimum value of P except as permitted in para. 122.1.3(A.4) shall exceed the maximum allowable working pressure of the boiler by either 25% or 225 psi, whichever is the lesser. (A.3) The value of P in the formula shall not be taken at less than 100 psig for any condition of service or material, and shall never be less than the pressure required to feed the boiler." This is one area that I have commonly seen overlooked.

From reference 7, paragraph "122.1.4 Blowoff and Blowdown Piping. (A.1) The value of P to be used in the formulas in para. 104 shall exceed the maximum allowable working pressure of the boiler by either 25% or 225 psi whichever is less, but shall be not less than 100 psig." This is one area that I have commonly seen overlooked.

From reference 7, paragraph "122.1.5 Boiler Drains. (D) Drain piping from the drain connection, including the required valve(s) or the blanked flange connection, shall be designed for the temperature and pressure of the drain connection. The remaining piping shall be designed for the expected maximum temperature and pressure. Static head and possible choked flow conditions shall be considered. In no case shall the design pressure and temperature be less than 100 psig and 220°F, respectively." This is one area that I have commonly seen overlooked.

From reference 7, see "NONMANDATORY APPENDIX II - RULES FOR THE DESIGN OF SAFETY VALVE INSTALLATIONS....II-2.3 Reaction Forces From Valve Discharge" and "II-3.0 BENDING MOMENT COMPUTATIONS". These are two areas that I have commonly seen overlooked.

From reference 7, see "NONMANDATORY APPENDIX VII - PROCEDURES FOR THE DESIGN OF RESTRAINED UNDERGROUND PIPING." This is one area that I have commonly seen overlooked.

13 Best Practice for Compressor Station Piping- ASME B31.3

From reference 4, paragraph "843.4.3 Air Piping System. (a) All air piping within gas compressing stations shall be constructed in accordance with ASME B31.3". This is one area that I have commonly seen overlooked.

From reference 4, paragraph "843.4.4 Lubricating Oil Piping. All lubricating oil piping within gas compressing stations shall be constructed in accordance with ASME B31.3." This is one area that I have commonly seen overlooked.

From reference 4, paragraph "843.4.7 Hydraulic Piping. All hydraulic power piping within gas compressing stations shall be constructed in accordance with ASME B31.3." This is one area that I have commonly seen overlooked.

So, to follow the best practice for compressor station air, lube oil and hydraulic piping, refer to reference 6, which has the following applicability "piping typically found in petroleum refineries; chemical, pharmaceutical, textile, paper, semiconductor, and cryogenic plants; and related processing plants and terminals."

From reference 6, see "NONMANDATORY APPENDIX II" for piping material information.

From reference 6, see "Table 302.3.4 Weld Quality Factor."

From reference 6, paragraph "304.1.1 General (a) The required thickness of straight sections of pipe shall be determined in accordance with $tm = t + c$, where

tm = minimum required thickness, including mechanical, corrosion, and erosion allowances
t = pressure design thickness, as calculated in accordance with para. 304.1.2 for internal pressure or as determined in accordance with para. 304.1.3 for external pressure
c = sum of the mechanical allowances (thread or groove depth) plus corrosion and erosion allowances. For threaded components, the nominal thread depth (dimension h of ASME B1.20.1, or equivalent) shall apply. For machined surfaces or grooves where the tolerance is not specified, the tolerance shall be assumed to be 0.5 mm (0.02 in.) in addition to the specified depth of the cut."

From reference 6, paragraph "304.1.2 Straight Pipe Under Internal Pressure
(a) For $t < D/6$, the internal pressure design thickness for straight pipe shall be not less than that calculated in accordance with eq. $t = PD / (2(SEW + PY))$ where
P = internal design gage pressure
D = outside diameter of pipe
S = stress value for material from Table A-1
E = quality factor from Table A-1A or A-1B
W = weld joint strength reduction factor in accordance with para. 302.3.5(e)
Y = coefficient from Table 304.1.1, valid for $t < D/6$ and for materials shown. The value of Y may be interpolated for intermediate temperatures. For $t \geq D/6$, $Y = (d + 2c) / (D + d + 2c)$ where d = inside diameter of pipe. For pressure design calculation, the inside diameter of the pipe is

the maximum value allowable under the purchase specification.
(b) For $t \geq D/6$ or for $P/SE > 0.385$, calculation of pressure design thickness for straight pipe requires special consideration of factors such as theory of failure, effects of fatigue, and thermal stress."

From reference 6, See "Table 314.2.1 Minimum Thickness of External Threaded Components." This is one area that I have commonly seen overlooked.

14 Best Practice for Piping – API 1102

API 1102, reference 9, is RAGAGEP that I have commonly seen overlooked.

From reference 9, "Table 2" is a good flowchart of the design process considering loads, stresses and fatigue.

From reference 9, paragraph "4.7.2.1 Stresses Due to Earth Load" has a calculation formula (1). Use Figures 3, 4 and 5 to get values to use in formula (1).

From reference 9, paragraph "4.7.2.2.1 Surface Live Loads", has guidance on rail road and highway road loads. Use equation (2) to determine the applied surface design pressure.

From reference 9, paragraph "4.7.2.2.2 Impact Factor" has guidance on impact factor to be used with live loads.

From reference 9, paragraph "4.7.2.2.3.1" has a has a calculation formula (3) for cyclic circumferential stress due to rail load. Use Figures 7, 8 and 9 to get the values to use in formula (3).

Reference 9, paragraph "4.7.2.2.3.2" has a has a calculation formula (4) for cyclic longitudinal stress for rail load. Use Figures 10, 11, 12 and 13 to get the values to use in formula (4).

Reference 9, paragraph "4.7.2.2.4.1" has a has a calculation formula (5) for cyclic circumferential stress for highway load. Use Table 2, Figures 14 and 15 to get the values to use in formula (5).

Reference 9, paragraph "4.7.2.2.4.2" has a has a calculation formula (6) for longitudinal circumferential stress for highway load. Use Figures 16 and 17 to get the values to use in formula (6).

Reference 9, paragraph "4.7.3" has a has a calculation formula (7) for stress from internal load.

Reference 9, paragraph "4.8.1.1" has a has a calculation formula (8a) to compare the circumferential stress due to internal gas pressure to the factored specified minimum yield strength. It also has a has a calculation formula (8b) to compare the circumferential stress due to internal liquid pressure to the factored specified minimum yield strength.

Reference 9, paragraph "4.8.1.2" has a has a calculation formula (9) to determine the maximum circumferential stress. It also has a calculation formula (10) to determine the maximum longitudinal stress. It also has a calculation formula (11) to determine the maximum radial stress. It also has a calculation formula (12) to determine the total effective stress. It also has a has a calculation formula (13) to compare the total effective stress to the factored specified minimum yield strength.

Reference 9, paragraph "4.8.2.1" has a has a calculation formula (14) to compare design to the girth weld fatigue. Table 3 is used with formula 14.

Reference 9, paragraph "4.8.2.1.1.1" has a has a calculation formula (15) to compare design to the girth weld fatigue for a railroad crossing for distance less than 5 feet.

Reference 9, paragraph "4.8.2.1.1.2" has a has a calculation formula (16) to compare design to the girth weld fatigue for a railroad crossing for distance 5 feet and greater. Figures 18A and B are used with formula 16.

Reference 9, paragraph "4.8.2.1.2" has a has a calculation formula (17) to compare design to the girth weld fatigue for a highway crossing.

Reference 9, paragraph "4.8.2.2.1" has a has a calculation formula (18) to compare design to the longitudinal weld fatigue.

Reference 9, paragraph "4.8.2.2.2" has a has a calculation formula (19) to compare design to the longitudinal weld fatigue for a railroad crossing.

Reference 9, paragraph "4.8.2.2.3" has a has a calculation formula (20) to compare design to the longitudinal weld fatigue for a highway crossing.

Reference 9, "Appendix B" has example design problems.

15 Best Practice for Piping – Guidelines for the Design of Buried Steel Pipe

Guidelines for the Design of Buried Steel Pipe, reference 10, is RAGAGEP that I have commonly seen overlooked.

Reference 10 has some additional useful information on piping design. Section "4.1" covers loads due to airport. Section "4.2" covers ovality, through wall bending, crushing of side walls and ring buckling. Section "5.1" covers maximum impact load. Section "5.2" covers penetration. Section "6" covers buoyancy. Section "11" covers earthquake. Section "12" covers nearby blasting. Section "13" covers fluid transients. "Appendix A" covers suggested acceptance criteria for stress and strain for 13 different loading conditions.

16 Conclusions

This paper presented design portions of CFRs for onshore metal pipelines. It guided the reader through the selection process of production/gathering and regulated/non-regulated in conjunction with API 80. It them went through design aspects of ASME B31.1/3/4/8, some of the equations and factors and the interrelationships between the ASME pipe codes, in particular for compressor stations. There are various loads and stresses that the CFRs and ASME pipe codes require to be considered, but many of the equations and methods are not detailed in those documents. Therefore, API 1102 and Guidelines for the Design of Buried Steel were discussed to fill in some of the gaps in the CFRs and ASME pipe codes, such as surface dead loads, surface live loads, and cyclic loads, failure mechanisms, equations, and acceptance criteria for deflection, ovality, through wall bending, crushing, ring buckling, fatigue, longitudinal, thermal expansion and fluid transients.

Using these presented methods results in inherently safer piping and safer pressure relief system design.

17 References

[1] API, RP 80, Guidance for the Definition of Onshore Gas Gathering Lines, 1st Edition, April 1, 2000, Reaffirmed January 2013.

[2] 49 CFR Part 192, Transportation of Natural and Other Gas By Pipeline: Minimum Federal Safety Standards, Revision 2/11.

[3] 49 CFR Part 195, Transportation of Hazardous Liquids by Pipeline, Revision 2/10.

[4] ASME, B31.8-2014, Gas Transmission and Distribution Piping Systems, ASME Code for Pressure Piping, B31, American Society of Mechanical Engineers, New York, NY, 2014.

[5] ASME, B31.4-2012, Pipeline Transportation Systems for Liquids and Slurries, ASME Code for Pressure Piping, B31, American Society of Mechanical Engineers, New York, NY, 2012.

[6] ASME, B31.3-2014, Process Piping, ASME Code for Pressure Piping, B31, American Society of Mechanical Engineers, New York, NY, 2015.

[7] ASME, B31.1-2004, Power Piping, ASME Code for Pressure Piping, B31, American Society of Mechanical Engineers, New York, NY, 2004.

[8] ASME, B1.20.1-2013, Pipe Threads, General Purpose (Inch), American Society of Mechanical Engineers, New York, NY, 2013.

[9] API, Recommended Practice 1102, Steel Pipelines Crossing Railroads and Highways, 6th Edition, American Petroleum Institute, Washington, D.C., April 1993, Reaffirmed July 2002.

[10] American Lifelines Alliance, Guidelines for the Design of Buried Steel Pipe, American Lifelines Alliance, July 2001.

13TH GLOBAL CONGRESS ON PROCESS SAFETY

Understanding Tolerable Risk Criteria – Considering the Growth of LNG Transportation

Ryan J. Hart, Ph.D., P.E.
Exponent, Inc.
4580 Weaver Parkway, Suite 100
Warrenville, IL
rhart@exponent.com

Delmar "Trey" Morrison, Ph.D., P.E.
Exponent, Inc.
4580 Weaver Parkway, Suite 100
Exponent, Inc.
Warrenville, IL
tmorrison@exponent.com

Prepared for Presentation at
American Institute of Chemical Engineers
2017 Spring Meeting and 13th Global Congress on Process Safety
San Antonio, TX
March 26 – 29, 2017

Understanding Tolerable Risk Criteria – Considering the Growth of LNG Transportation

Ryan J. Hart, Ph.D., P.E.
Exponent, Inc.
4580 Weaver Parkway, Suite 100
rhart@exponent.com

Delmar "Trey" Morrison, Ph.D., P.E.
Exponent, Inc.
tmorrison@exponent.com

Keywords: risk assessment; tolerable risk criteria; transportation risk

Abstract

Cheap natural gas and the aim to lower emissions from combustion systems such as stationary power generation and heavy haul vehicle engines has led to a multi-industry effort to convert to natural gas equipment. Natural gas pipelines can reach most of the U.S., but there are many areas where shipping liquefied natural gas (LNG) via truck or train may be the most economical solution. This article focuses on shipping LNG via rail and benchmarking the risk this poses. Although similar to other hazardous materials, LNG shipping via rail is a new endeavor in the U.S., and the regulations do not currently allow this to happen. The Federal Rail Administration (FRA) is the government entity that regulates rail shipping, and has implemented a requirement for a risk-based analysis in order to get approval to ship high hazard flammable trains (HHFT) after a series of crude oil train accidents over the last decade. Further, the FRA has issued specific guidance related to risk assessment for LNG trains. Quantitative Risk Assessment (QRA) is one tool that can be applied to evaluate the risks posed by shipping LNG. This falls into a special class of QRA called Transportation Risk Analysis.

But how does one evaluate the quantified risk for shipping LNG? How can it be demonstrated that the risk of shipping LNG is acceptable to society? Currently, the U.S. federal government has not explicitly provided risk criteria for evaluating acceptability of shipping LNG via specific routes. In contrast, international regulators and industry organizations have implemented tolerable risk criteria for determining acceptability of new technologies, but their underlying bases may not be apparent. In this paper, we will demonstrate how existing stationary LNG plant tolerable risk criteria can be used to develop criteria for LNG transportation risk. The inappropriateness of directly applying these existing stationary plant risk criteria to transportation will be explained, and another existing HHFT commodity (propane (LPG) for example) will be used to benchmark against the stationary LNG plant criteria and to show how these criteria can be modified to apply to transportation risk. Ultimately, the risks will be quantified for LNG HHFTs and compared to existing HHFTs to show that the newer fuel, LNG, poses reasonable and equivalent risks to many of the other HHFTs currently shipped through the U.S. From this work, a tolerable risk criterion for LNG shipping via rail will be proposed.

1 Introduction

The growth in domestic natural gas production is expected to continue to grow over the next several decades (Figure 1) [1]. The surplus of natural gas is expected to be handled through exports as well as domestic consumption. More and more we will start seeing LNG transport containers being shipped over the road and along the rail. To what extent is yet to be determined, but an increase in transportation of LNG within the U.S. is an almost certainty.

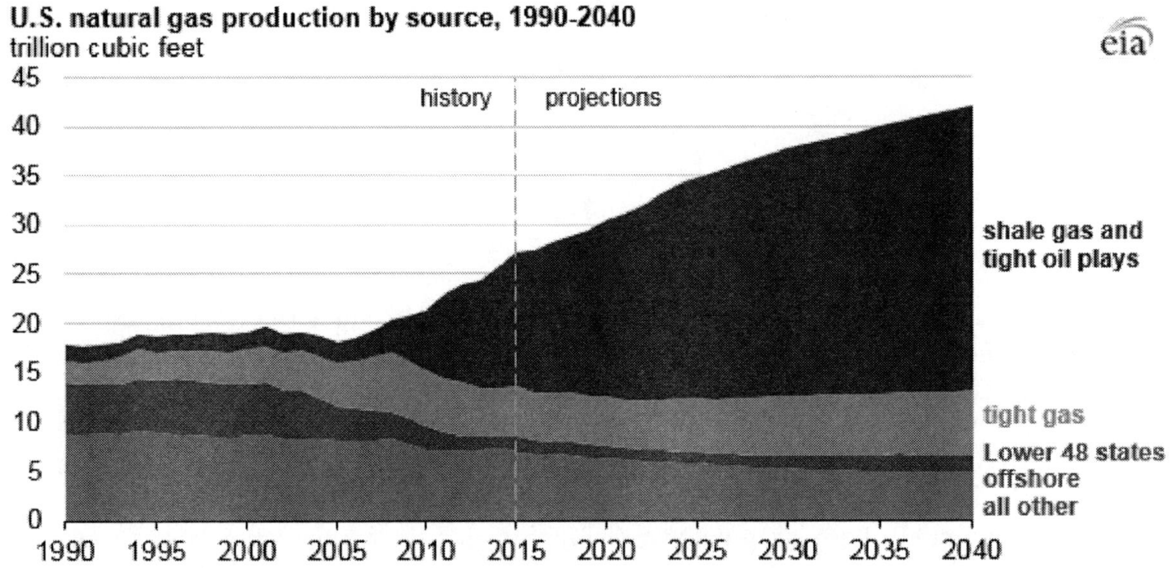

Figure 1. Project domestic natural gas production [1].

Although LNG transportation has existed for decades within the U.S., the anticipated surge in transportation is not without concern. Given the high visibility of transportation incidents involving crude oil, much attention is being paid to LNG transport by rail. Ensuring that the development and implementation of LNG transportation is performed safely is of critical importance to industry, communities, and regulators.

Much of the concern surrounding LNG transport may be attributed to a general lack of understanding regarding the unique and particular hazards of LNG. Quantitative risk assessment (QRA) is a tool that has been used successfully in the past to evaluate the potential risks of emerging technologies and proposed transportation routes. QRA is not a common public-policy tool in the U.S., but it is beginning to grow in popularity with private companies, regulatory agencies, and standards committees alike. Most QRA has been performed for stationary facilities, which are inherently different than transportation projects. In this paper, we discuss tolerable risk criteria for use in evaluating the QRA of transport projects. We anticipate that this approach may be useful for evaluating the blossoming LNG transport in the U.S.

2 Understanding Quantitative Risk Assessments

Risk, simply defined, is the potential to lose something of value. Risk is calculated by taking the product of event likelihood with the event outcome severity. A simplified flow diagram depicting the process for determining risk is provided in Figure 2.

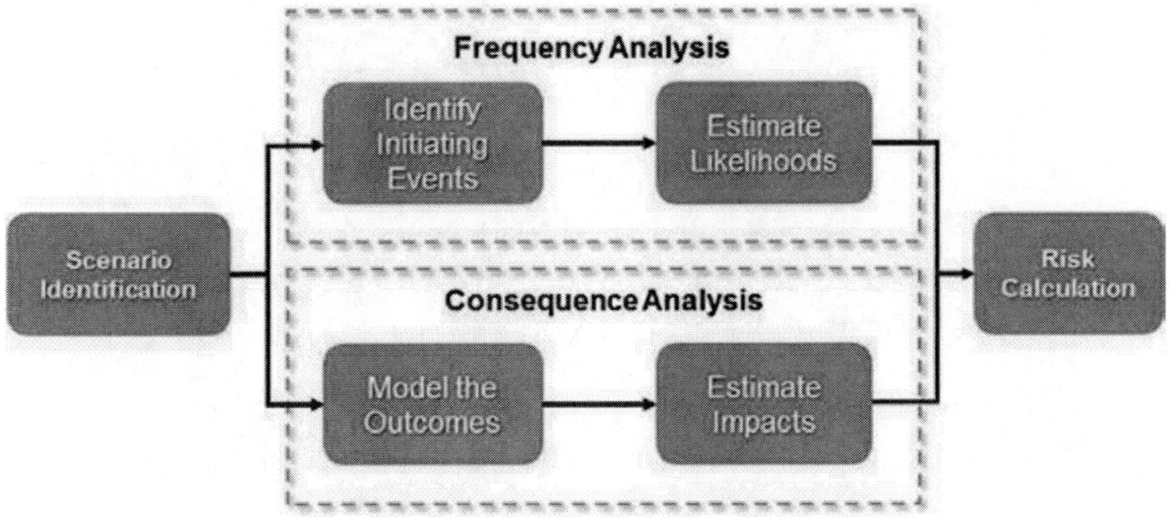

Figure 2. Flow diagram depicting the process for calculating risk.

The likelihood of an event can be estimated using experience relating to given equipment in similar service, industry data, or engineering approximations. A challenge of quantifying risk, or affixing a number to a particular risk level, is determining how to quantify the event outcome portion of the equation. For quantifying risk at industrial facilities and transportation operations, the outcome of an event is typically evaluated as the potential for a fatality or multiple fatalities.

In evaluating the potential for fatality, two metrics are utilized to yield the risk: (1) Individual Risk (IR) and (2) Societal Risk (SR). IR is the frequency (yr^{-1}) where an individual with continuous potential exposure may be expected to sustain a serious or fatal injury. When calculating IR, it is assumed that persons are constantly occupying all space where an incident may express an undesired impact. As such, IR is largely independent of population density except to the extent that the likelihood component of the risk is dependent on population (e.g. likelihood of ignition calculation). IR results are typically presented as frequency contours on aerial maps that illustrate the risk to individuals positioned within those contours. An example of an IR result is presented in Figure 3.

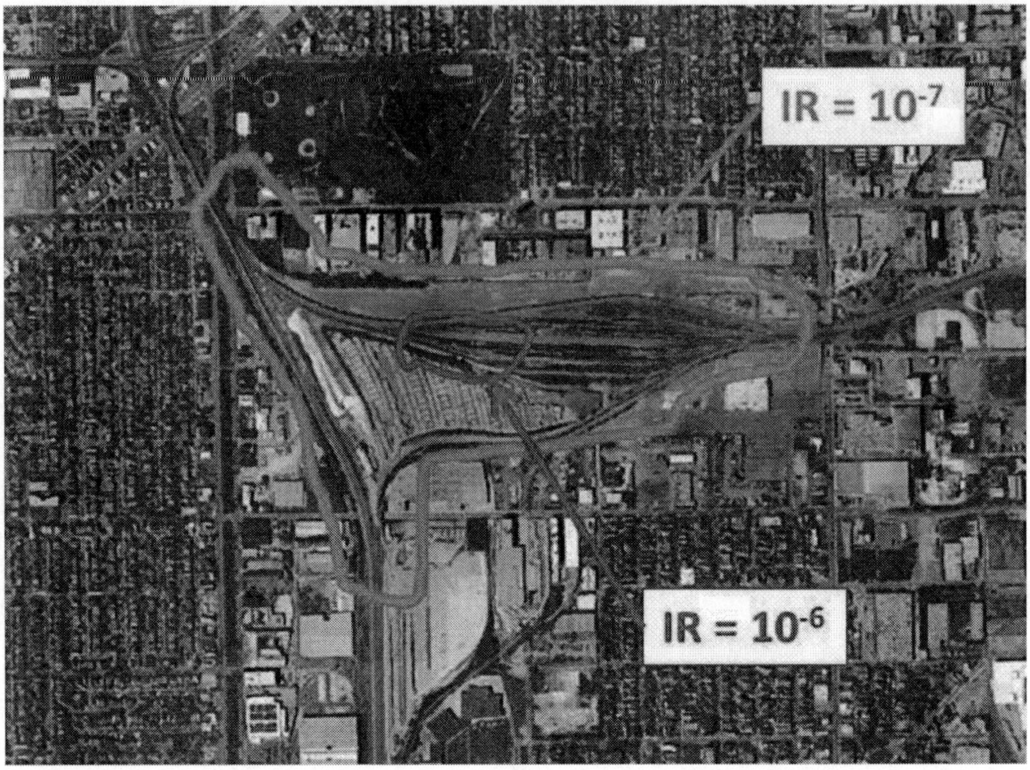

Figure 3. Example of an IR result, presented around a rail yard.

Societal Risk (SR) is another metric for evaluating the risk of a given facility or operation. Unlike IR, the SR calculation considers the relationship between the cumulative number of potential fatalities and the likelihood from a series of potential events. The outcome of a SR analysis is a graph depicting annual frequency on the y-axis and N fatalities on the x-axis, where N is the cumulative number of potential fatalities for all scenarios represented by the corresponding cumulative frequency of events. Unlike IR, the SR calculation takes into account the differences in population density in the vicinity surrounding the incident. Whereas the IR calculation gives insight into the probability of having a fatality at a specific location, the SR calculation gives the likelihood of a number of potential fatalities covering the entire operation. This is especially important for evaluating scenarios with a large potential impact for loss of life, such as train derailments of flammable materials. An example of an SR result for two separate scenarios is provided in Figure 4.

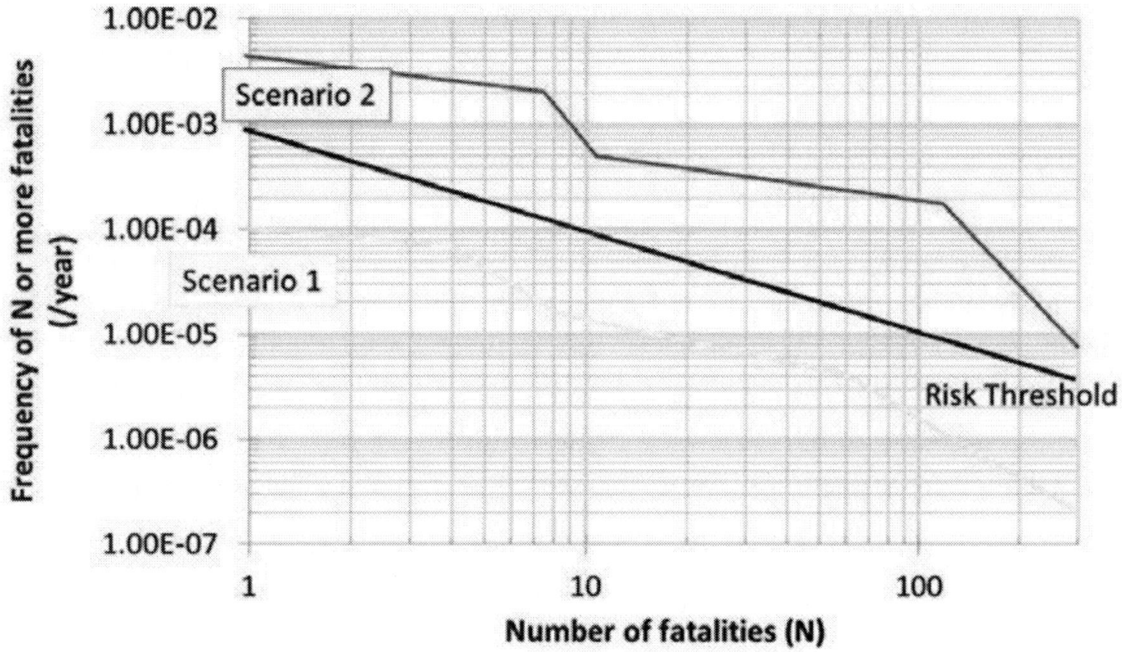

Figure 4. Example of an SR result for two different scenarios.

Quantitative risk assessments may be performed for several reasons:

1. Comparing the relative risk of two or more competing options (e.g. comparing route options for transporting hazardous materials)
2. Identifying high risk aspects of a specific operation
3. Demonstrating that the risks of a certain operation are tolerable or acceptable

When analyzing risks to demonstrate their acceptability or tolerability, the selection of appropriate risk tolerability criteria is paramount. Selection of inappropriate tolerability criteria could potentially result in a false-confidence in the perceived safety (in instances where the risk tolerability may be too high) or unnecessary elimination of a proposed project or operation (in cases where the risk tolerability may be too low).

3 Quantitative Risk Tolerability Criteria

After quantifying risk and presenting the calculations as IR and SR for a given operation or process, the results are evaluated for tolerability (or acceptability). Typically, stakeholders (e.g., government agencies, investors, communities) have a threshold risk level that is deemed acceptable—known as quantitative risk criteria. Currently, the U.S. federal regulatory agencies have not codified quantitative risk criteria for hazardous materials transportation scenarios [2] and QRA analyses are not common regulatory requirements in the U.S. Several foreign regulatory agencies, such as the Dutch government, have been international leaders in utilizing QRA techniques for determining acceptability of fixed facilities and transportation routes.

However, no broadly-accepted risk criteria are employed by domestic communities or industries. There are several foreign and several domestic examples of quantitative risk criteria (see [3]-[5]) and these will be summarized in the next sections for both IR and SR.

Recommendations for QRA of LNG plants were issued in the National Fire Protection Association (NFPA) standard, NFPA 59A *Standard for the Production, Storage, and Handling of Liquefied Natural Gas (LNG)* [6]. In addition to including QRA as a risk assessment tool in NFPA 59A, the standard also includes quantitative risk criteria for fixed LNG facilities. NFPA 59A explicitly applies to LNG plants and stationary facilities; it does not apply to LNG transportation or portable LNG containers. Thus, the tolerable quantitative risk criteria proposed in the standard are not directly applicable to the transportation of LNG, as will be discussed below.

There are significant differences between stationary facilities and transportation projects, for example:

- Size scale of the enterprise. The risk calculation is directly related to the number of vessels or components and the length of piping systems or a route. In a fixed facility, the length scale is fixed and on the order of hundreds to thousands of feet. In transport, the length scale may be hundreds to thousands of miles. Thus, transport is anticipated to create higher risks for similar hazmat.
- Site location. Stationary facilities that are yet to be built typically have the opportunity to survey multiple potential sites, whereas transportation routes options are more restricted (in some cases, only one transportation route may be an option).
- Siting. For stationary facilities, consideration can be given where on the site to place hazardous material operations in order to minimize potential off-site impacts or reduce the likelihood of suffering cascading events. Transportation of hazardous materials has little flexibility in placement along a given route and has no control over adjacent hazards (e.g. a hazmat transport carrier cannot prevent another hazmat transport carrier from overtaking them on the highway).
- Standoff distances. Proposed facilities may have the opportunity to acquire more land for a given project in order to increase the distance from hazmat operations to the property boundary. The separation distances between a transport carrier and the public are already established due to the zoning of the transport route and adjacent properties. Additionally, in some transport modes (like over the road trucking) the transport carrier has no choice but to be a few feet from a public receptor.
- Control of ignition sources. Facilities operating with hazmat have complete control over the abundance and location of ignition sources on their site. Transport routes have no control over ignition sources, and nearly every public transporter can be assumed to be a competent ignition source.
- Training/emergency response. Facilities have the opportunity to train personnel and prepare for an onsite release of a hazmat. Transport carriers may have a highly trained crew for their specific carrier, but the public in the immediate vicinity of the hazard is presumed to have no training with respect to the hazards presented by hazmat. Additionally, hazmat transport relies upon the emergency response of the jurisdiction through which it travels, which may change dozens or hundreds of times on a given

route. Facilities, by comparison, usually have only 1 or 2 emergency responder agencies to coordinate with.

For the reasons stated above, the quantitative risk criteria for stationary facilities should be different than the quantitative risk criteria for transportation routes. The risks to the onsite and offsite receptors associated with transporting hazmat are inherently different than the risks to those populations surrounding a stationary facility. To account for these differences, transportation risks should be evaluated differently than stationary facility risks.

3.1 Individual Risk Criteria

Among the available references for quantitative risk criteria used domestically and internationally, there is a wide disparity in risk criteria for public exposure, with acceptable IR fatality probabilities ranging from 10^{-4} yr^{-1} (or a fatality per 10,000 years) to 10^{-8} yr^{-1} (or a fatality per 100,000,000 years). NFPA 59A identifies three "Zones" representing ranges of quantitative risk criteria for evaluating IR at stationary LNG facilities. Each risk zone reflects general types of public occupancies recommended to be permitted. As the magnitude of the calculated risk increases, the type of occupancy becomes more restrictive. The quantitative risk criteria for IR of LNG plants, according to NFPA 59A, are reproduced in Table 1.

Based on Zone 3 being the most restrictive zone, any IR values less than 3×10^{-7} yr^{-1} are not of concern for the analysis. The IR ranges and associated criteria reported in NFPA 59A appear to be based on guidance provided by the Health and Safety Executive in the U.K. for QRA [7] and do not account for the factors typically considered in a transportation risk analysis. Conversely, a commonly acceptable level of IR for transportation risks on sensitive populations is 10^{-6}, which is the upper threshold for Zone 3 [5].

Table 1. Quantitative risk criteria for IR contours as provided by NFPA 59A (2016).

Criterion Annual Frequency (yr^{-1})	Remarks
Zone 1 $IR > 10^{-5}$	Not permitted: Residential, office, and retail Permitted: Occasionally occupied developments (e.g., pump houses, transformer stations)
Zone 2 $10^{-6} \leq IR < 10^{-5}$	Not permitted: Shopping centers, large-scale retail outlets, restaurants, etc. Permitted: Work places, retail and ancillary services, residences in areas of 7,250 to 23,300 persons/mile^2 density
Zone 3 $3 \times 10^{-7} \leq IR < 10^{-6}$	Not permitted: Churches, schools, hospitals, major public assembly areas, and other sensitive establishments Permitted: All other structures and activities

For flammable material release scenarios, the magnitude of the risk generally increases as the observation point is moved closer to the transport route. The distance to each risk level identified in the table is a result of the compilation of the outcomes calculated from an event tree of many potential fire and explosion events. An example of calculated distances to various IR thresholds for the transportation of LNG and LPG have been reported **Error! Reference source not found.**, reproduced in Table 2. LPG represents an inherently acceptable hazmat commodity, which has been shipped on railroads for a long time. Table 2 shows that the IR of LNG is similar or less than LPG for both road and rail transport.

Table 2. Example calculated IR distances for the road and rail transport of LNG and LPG.

				IR Distance to...	
Material	**Transport Mode**	**Pop Density (/mile2)**	**10^{-5} IR (m)**	**10^{-6} IR (m)**	**$3x10^{-7}$ IR (m)**
LPG	Road	10,000		118	177
		1,000		86	115
		100		84	108
LNG	Road	10,000		119	169
		1,000		62	95
		100		56	84
LPG	Rail	10,000	186	290	334
		1,000	178	290	334
		100	174	290	334
LNG	Rail	10,000	92	193	213
		1,000		145	184
		100		104	132

It should be noted that the analysis in **Error! Reference source not found.** was conservative, especially in the instances of rail risks where it was assumed that a train carrying a hazmat would involve the hazmat rail car in all instances. Despite the conservativeness, the data illustrates that the NFPA 59A Zone 3 criteria for IR (3×10^{-7} yr^{-1}) is much more restrictive than the Zone 2 criteria (1×10^{-6} yr^{-1}). Additionally, the use of the Zone 3 criteria would indicate that the inherently acceptable risks of transporting LPG may not satisfy this criterion in instances where there are sensitive targets within 334 m of railroads and 177 m of roads where LPG is transported.

3.2 Societal Risk Criteria

As with IR, the international and domestic quantitative risk criteria for SR can vary greatly. Most SR criteria present two tolerability lines with three distinct regions for which to evaluate the resulting FN curve. For discussion purposes, the SR criteria from NFPA 59A is provided in Figure 5. There are two lines presented in the figure:

(1) an <u>upper risk criteria line</u> (solid red line), commonly referred to as the maximum risk line, and

(2) a lower risk criteria line (dashed line), commonly referred to as the tolerable risk line.

For calculated FN curves that fall above the upper risk criteria line, the risk is considered "unacceptable." Calculated FN curves that fall below the lower risk criteria line are determined to be acceptable, and FN curves between the two criteria lines fall within the "As Low As Reasonably Practicable (ALARP)" region. Risks falling within the ALARP region are typically evaluated in more detail for opportunities to either mitigate or reduce the perceived risk of the operation, in the form of additional engineering or administrative controls.

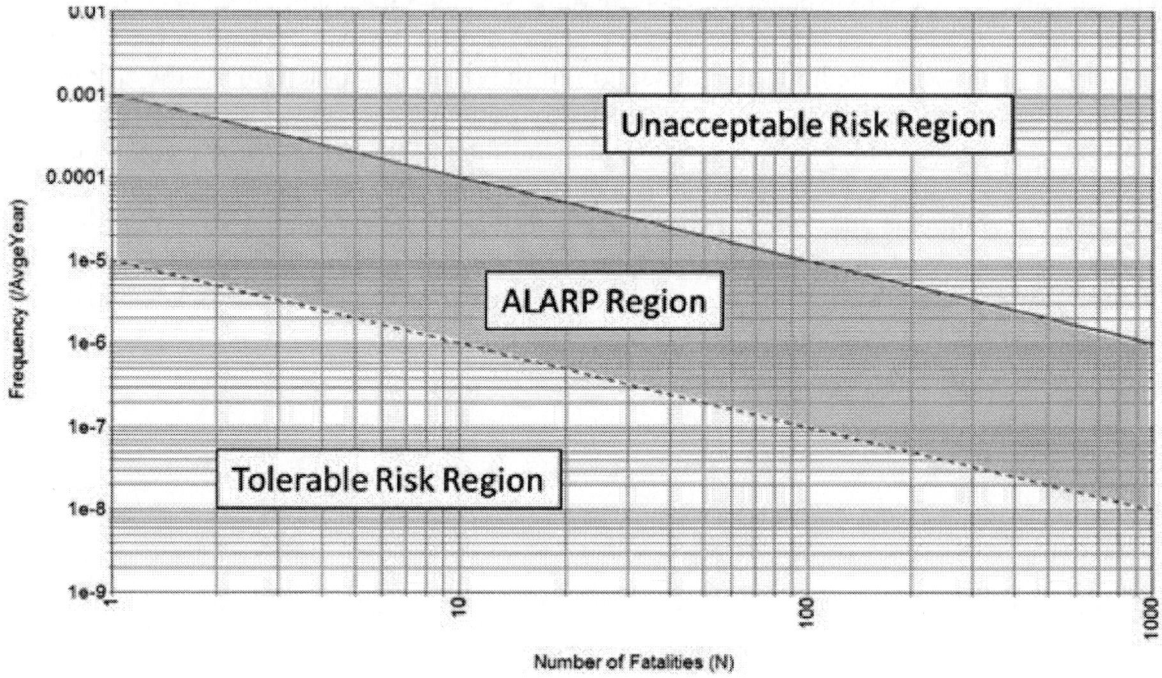

Figure 5. NFPA 59A (2016) SR tolerability criteria.

Most international and domestic examples of SR criteria follow the same general format with both an upper and lower risk bound. Although different regulatory agencies, standards committees, and private companies may have different definitions of upper and lower risk lines, nearly all examples of SR criteria apply to only stationary facilities.

Using a previous reported QRA study for the transport of LNG and LPG [8], we have adopted the data here to demonstrate calculated FN curves. Figure 6 depicts the calculated FN curves for the rail and road transport of LNG and LPG along a 50-mile long route surrounded by a population with a density of 1,000 people/mile^2. The figure also includes the maximum and minimum risk lines from NFPA 59A, and the ALARP region highlighted in grey. The figure illustrates the potential issues when directly applying SR criteria intended for stationary facilities to transportation risks. The risks for all four transport routes fall within the "Unacceptable Risk" region as defined in NFPA 59A. LPG is certainly transported by truck and rail through similar distances and population densities, but the risk to society is acceptable.

Although most SR criteria available both domestically and internationally are applied to stationary facilities, methods have been developed to apply the stationary SR criteria to transportation routes. These methods involve normalizing the scale of the transport project risks and modifying the risk criteria lines. To correct the fixed facility SR criteria for application to a transportation project that may be considerably larger in scale, the SR for transportation routes are calculated for a fixed unit of length. This allows for similar hazard scales to be compared between transport routes and fixed facilities. The common unit for normalizing the FN curves of transportation projects is either 1 mile or 1 km, depending on the common engineering units in the project's location. This can be accomplished by either (1) calculating the FN curve for the entire transportation route and dividing the y-axis (frequency) by the route length or (2) conservatively assuming that the highest-risk section of the transport route bounds the risk along the entire line. Such an approach is sensible because any potential incident can only affect a limited portion of the society along the route for these flammable materials; thus, using a per-mile or per-kilometer basis for SR is reasonable. The same approach may not be acceptable for toxic materials, e.g., chlorine gas, which could project significant hazards over a greater distance. An additional approach is to compare the aggregate SR curve for the route for the new hazmat (LNG) to an existing hazmat (LPG). If the new curve is similar or less, then it poses less risk along the route than the reference material.

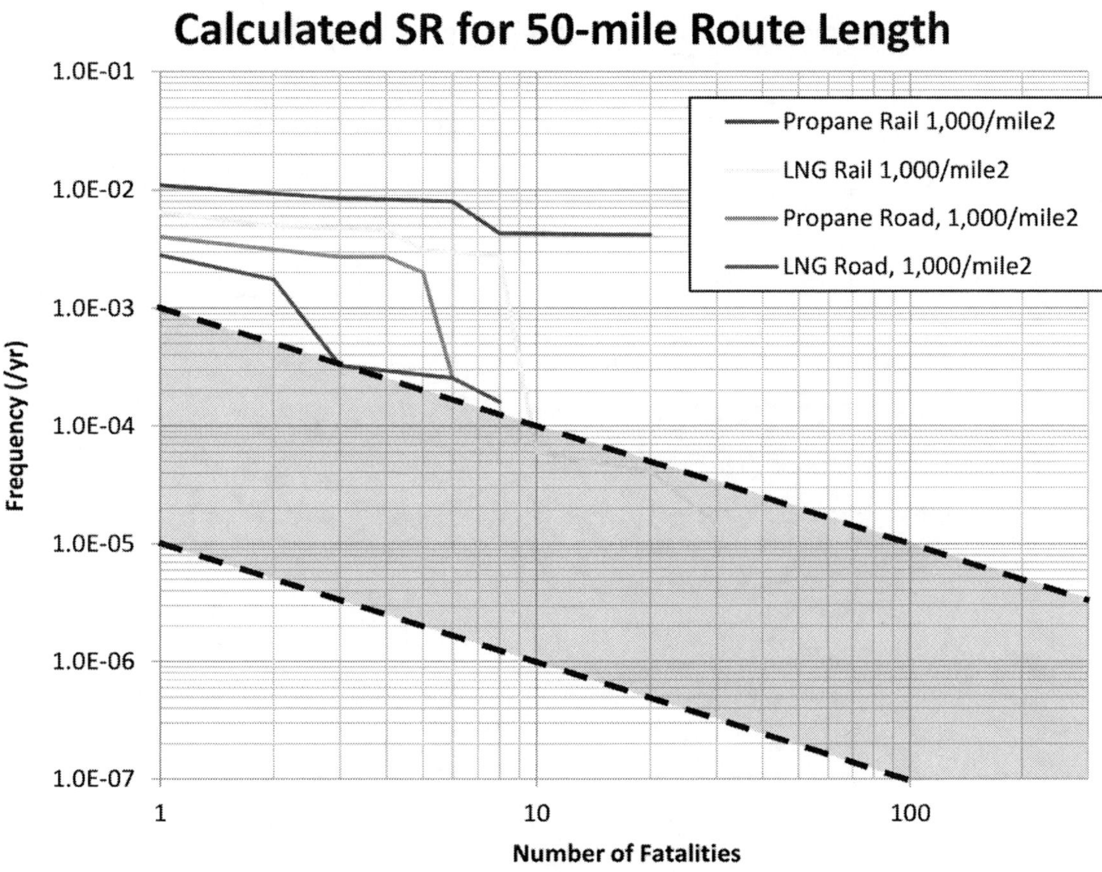

Figure 6. Calculated FN curves for the road and rail transport of LNG and LPG along a 1-mile long route surrounded by a population with a density of 1,000 people/mile^2.

Figure 7 provides the FN curves for the same transport routes above, but normalized on a 1-mile length. All other inputs are identical between the two graphs. When scaling the transportation FN curves to be comparable to the length scale of stationary facilities, the application of the stationary facility SR criteria becomes more reasonable. However, it can be seen that one of the analyzed transportation routes, propane rail transport, still exhibits a portion of the FN curve with the NFPA 59A defined "unacceptable risk" region. As previously mentioned, there are other considerations between project scales that distinguish the risks of stationary facilities from transportation projects. In order to consider for these other differences (such as ignition source control, training, siting), some international organizations increase the maximum and minimum risk lines by an order of magnitude – the implication here is that the tolerability of risks associated with transportation are inherently higher that for exposure to a stationary facility. If this approach were to be taken for the FN curves presented in Figure 7, none of the FN curves would lie within the "unacceptable risk" region (in fact, the LNG Road scenario would lie in the "acceptable risk" region).

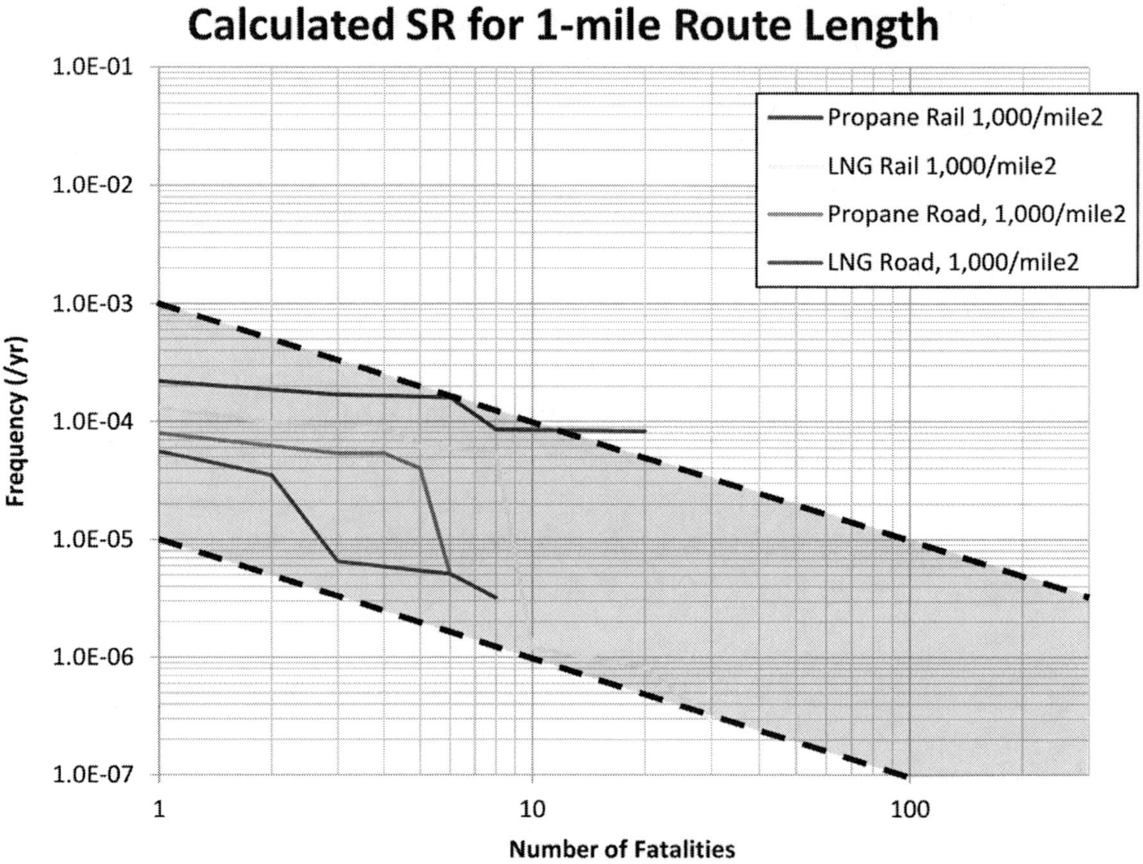

Figure 7. Calculated FN curves for the road and rail transport of LNG and LPG along a 1-mile long route surrounded by a population with a density of 1,000 people/mile^2.

4 Conclusions

Quantitative risk assessments are powerful tools for assessing the risks of emerging and existing hazardous material operations within stationary facilities and along transportation routes. The risks can be assessed through either comparative analyses or by application of tolerable risk criteria. Many international and domestic organizations, both public and private, have identified tolerable risk criteria for evaluating both Individual Risk (IR) and Societal Risk (SR), but considerable variability exists. Additionally, most tolerable risk criteria have been developed for stationary facilities.

As shown in this paper, application of risk criteria developed for stationary facilities to transportation routes may not be appropriate. The risks of transporting hazardous materials is different from fixed facilities, as are the thresholds for risk tolerability that society is inherently willing to accept. The application of improper risk criteria may prevent or delay the proposed transportation of emerging materials, such as LNG.

Based on previously reported IR and SR values for LNG and LPG transport, we have demonstrated that although the risks of LNG transport are less than the risks of transporting LPG, certain risk tolerability criteria may indicate that the risks are unacceptable. This conflicts with the inherently acceptable risks of LPG transport, which were demonstrated to be higher than those for comparable LNG transport.

A survey of domestic and international risk tolerability was performed. Most risk tolerability criteria have been developed for and applied to stationary facilities. The risks of stationary facilities are inherently different that transportation risks for two primary reasons: (1) transportation routes can be significantly larger than stationary facilities and (2) the inherent tolerability of transportation risks is different than for stationary facilities. Based on the analysis of representative LNG and LPG transport risks, compounded with the typical treatment of risk tolerability criteria by international organizations, methods for using stationary facility risk criteria to evaluate transport risks are proposed:

- IR Tolerability Criteria – for transport projects, the use of 10^{-6} yr^{-1} as the IR frequency threshold for impacts to sensitive targets is consistent with an abundance of international sources. This value is approximately one order-of-magnitude higher than the minimum sensitive target tolerability criteria presented in NFPA 59A for stationary LNG facilities (3×10^{-7} yr^{-1}).
- SR Tolerability Criteria – for transport projects, normalizing the FN curves for a representative length (1 mile or 1 km) can allow for re-scaling of transportation risks to a stationary facility. Further, to account for the higher risk tolerability for transportation relative to stationary facilities, the maximum and minimum risk criteria lines may be increased by an order of magnitude.

5 References

[1] U.S. Energy Information Administration, "Annual Energy Outlook 2016," July 7, 2016.

[2] Strang J, "Federal Railroad Administration Risk Reduction Programs," United States Army Corps of Engineers Workshop on Tolerable Risk, March 18-19, 2008, Alexandria, Virginia.

[3] Appendix B: Survey of Worldwide Risk Criteria Applications, *Guidelines for Developing Quantitative Safety Risk Criteria*. Center for Chemical Process Safety, AIChE (2009).

[4] Cornwell JB and MM Meyer, "Risk Acceptance Criteria or 'How Safe is Safe Enough?'" presented at II Risk Control Seminar in Puerto La Cruz, Venezuela, October 13, 1997.

[5] Ham JM, M Struckl, AM Heikkila, E Krausmann, C DiMauro, M Christou, JP Nordvik, "Comparison of Risk Analysis Methods and Development of a Template for Risk Characterisation," Institute for the Protection and Security of the Citizen, European Commission, Directorate-General Joint Research Center (2006).

[6] NFPA 59A, Standard for the Production, Storage, and Handling of Liquefied Natural Gas (LNG), 2016 edition, National Fire Protection Association.

[7] "B.1 Evolution of Land Use Planning Criteria in the UK," in Guidelines for Developing Quantitative Safety Risk Criteria, American Institute of Chemical Engineers, Center for Chemical Process Safety (2009).

[8] Hart RJ, Morrison DR. The hazard we know: Comparing transportation risk of LPG and LNG. American Institute of Chemical Engineers, 2015 Spring National Meeting and 11th Global Congress on Process Safety, Austin, TX, April 26-30, 2015.

Challenges in the Application of Adsorbent Technologies for Midstream Contaminants Control

Mahdi Nouri, CEng FIChemE
CH2M

Paul Clewis, Olivier Ducreux, Lorrel Hodgson
Axens North America

Table of Contents

Table of Contents

- **Basics of Adsorption**
- Drying Challenges and Solutions
- Mercury Removal Challenges and Solutions
- Sulfur Compound Removal Challenges and Solutions
- Conclusions

17AIChE
Spring Meeting, San Antonio, TX

3

ch2m: Axens IFP Group Technologies

Adsorbents

- Adsorbents are materials used to remove undesirable contaminants from liquids and gases
- Adsorbents include:
  1. Activated Alumina
  2. Promoted Alumina
  3. Traps/Scavengers
  4. Molecular Sieve

- Adsorbents purify streams for equipment protection and product specifications

17AIChE
Spring Meeting, San Antonio, TX

4

ch2m: Axens IFP Group Technologies

Two Types of Adsorption

- There are two types of adsorption:
  1. Physical adsorption (physisorption)
  2. Chemical adsorption (chemiosorption)

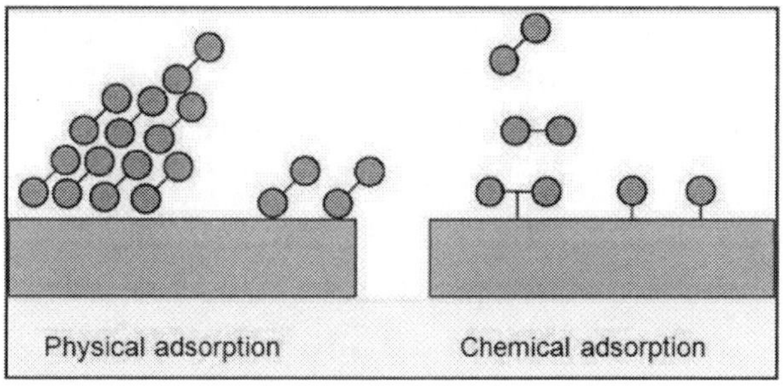

- The main differences come from
  1. the nature of the bond between the adsorbent and impurity
  2. the energy level involved
- Sometimes there is no clear distinction

Physical Adsorption

- Also called Physisorption, it is a process in which the electronic structure of the atom or molecule is barely perturbed upon adsorption, and generally remains practically unchanged. It is a general phenomenon that can occur in any solid/fluid or solid/gas system. The process is generally driven by only electrostatic and van der Waals interactions, and no chemical bonding is involved.

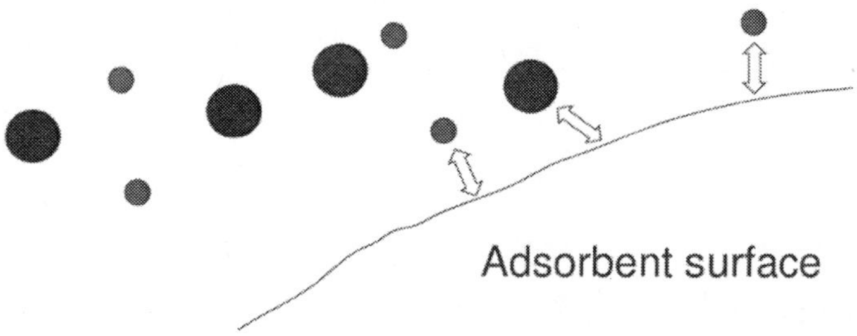

Chemical Adsorption

- Also called Chemisorption, it is a kind of adsorption which involves a chemical reaction between the surface and the adsorbate, with new chemical bonds being generated at the adsorbent surface. The term can actually describe a wide range of chemical activities, but with the distinction that a strong interaction between adsorbate and substrate results in the creation of new, irreversible bonds.

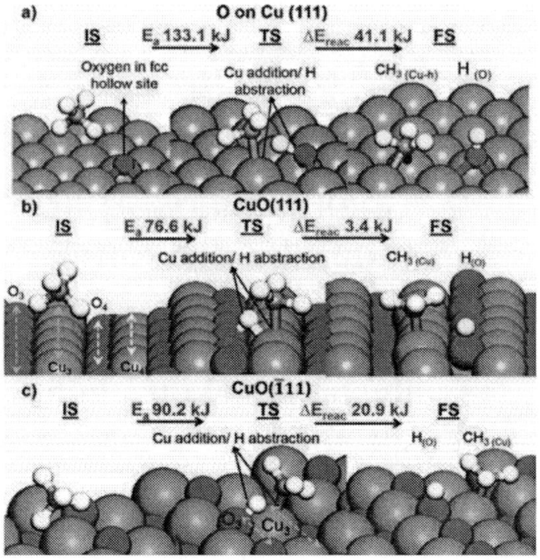

Natural Gas Contaminants

- Water (H_2O)
- Elemental Mercury (Hg^0)
- Hydrogen Sulfide (H_2S)
- Mercaptans (RSH)
- Carbonyl Sulfide (COS)
- Carbon Disulfide (CS_2)
- Carbon Dioxide (CO_2)

Natural gas is water saturated and depending on the world area is more and less acidic and contains more and less elemental mercury

Natural Gas Contaminants

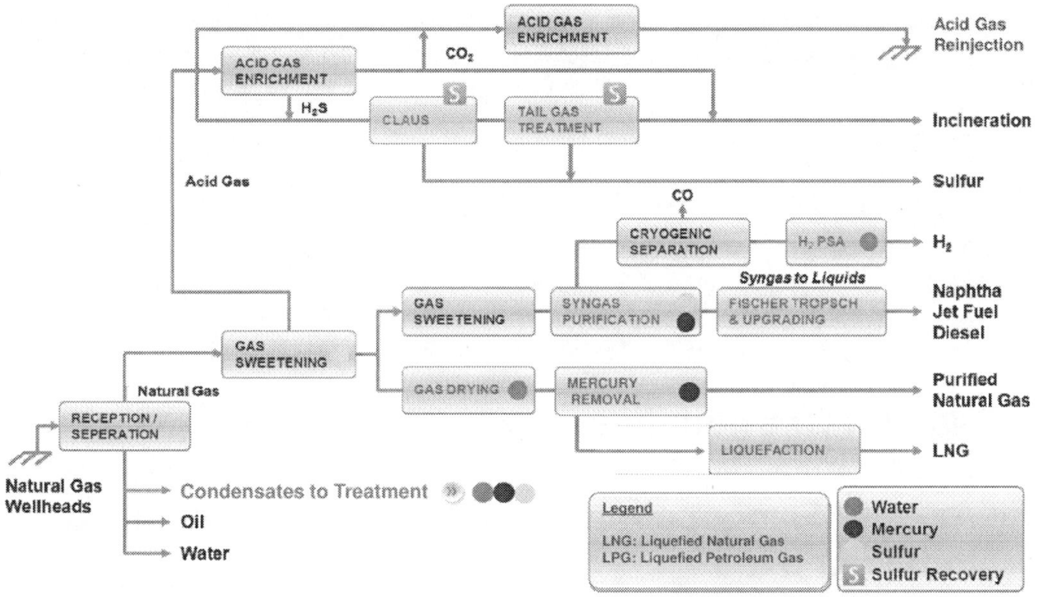

17 AIChE Spring Meeting, San Antonio, TX

9

ch2m: Axens IFP Group Technologies

Natural Gas Contaminants

Contaminants	Main Reasons of Removal
Water (H_2O)	Freezing and equipment plugging
Elemental Mercury (Hg^0)	Corrosion and toxicity
Hydrogen Sulfide (H_2S)	Corrosion
Mercaptans (RSH)	Corrosion
Carbonyl Sulfide (COS)	Corrosion
Carbon Disulfide (CS_2)	Corrosion
Carbon Dioxide (CO_2)	Freezing and equipment plugging

The degree of removal depends on downstream requirements

17 AIChE Spring Meeting, San Antonio, TX

10

ch2m: Axens IFP Group Technologies

Specifications

Contaminants	Pipeline Specs	LNG/NGL Specs
H_2O	7 lbs/MMSCF	0.01 ppmv
Hg^0	-	0.01 μg/Nm^3
H_2S	10-15 mg/Nm^3	5 mg/Nm^3
Total Sulfur	70 mg/Nm^3	30 mg/Nm^3
CO_2	2 vol%*	50 ppmv

*In order to increase calorific value

- **Natural Gas liquefies at -256 °F (-160°C)**
- **At -256 °F, most contaminants are solid and need to be removed**

Carbon Dioxide Removal Options

- **Carbon dioxide removal to low ppm levels can be accomplished by a promoted alumina, molecular sieve, or hybrid of both products**

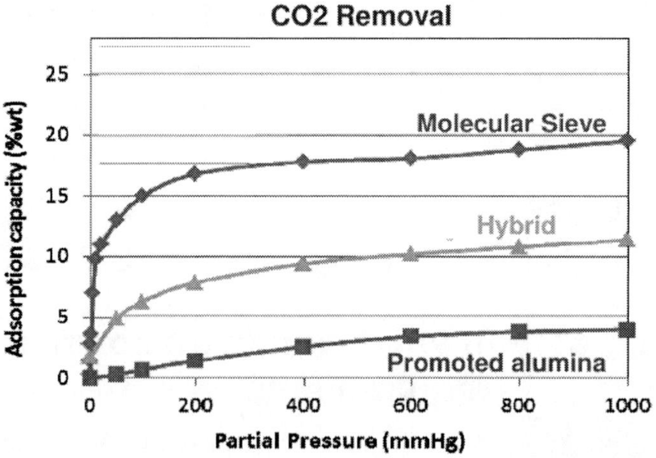

Activated Alumina

- Activated alumina- a porous form of Al_2O_3 with high surface area
- Commonly used for:
  - Drying
- Advantages:
  - Regenerable
  - Robust material with no binder
  - No olefin co-adsorption

17 AIChE Spring Meeting, San Antonio, TX — 13 — ch2m: Axens IFP Group Technologies

Promoted Alumina

- Promoted alumina- a porous form of Al_2O_3 with high surface area and promoters to modify its surface functionality
- Commonly used for:
  - Acidic molecule removal like COS, CS_2, and CO_2
  - Advantages:
  - Enhanced contaminant trapping capacity and selectivity
  - Can treat contaminants to low outlet specifications
  - Regenerability based on service
  - Robust material with no binder
  - No olefin co-adsorption

17 AIChE Spring Meeting, San Antonio, TX — 14 — ch2m: Axens IFP Group Technologies

Traps & Scavengers

- **Definition- non-regenerable material used for irreversibly trapping impurities**
- Commonly used for:
  - Mercury removal, H_2S removal, RSH removal, and COS removal
- Advantages:
  - Strong chemisorption mechanism
  - Kinetically favored reactions with contaminants
  - Can treat contaminants to low outlet specifications

17 AIChE Spring Meeting, San Antonio, TX

15

ch2m. Axens IFP Group Technologies

Molecular Sieve

- **Molecular sieve- aka zeolite; aluminum silicate crystalline material with ~20 wt% base cations**
- Commonly used for:
  - Drying, H_2S removal, RSH removal, and CO_2 removal
- Advantages:
  - High water removal capacity at low partial pressure
  - Different micropore sizes capture different contaminants
  - Regenerable

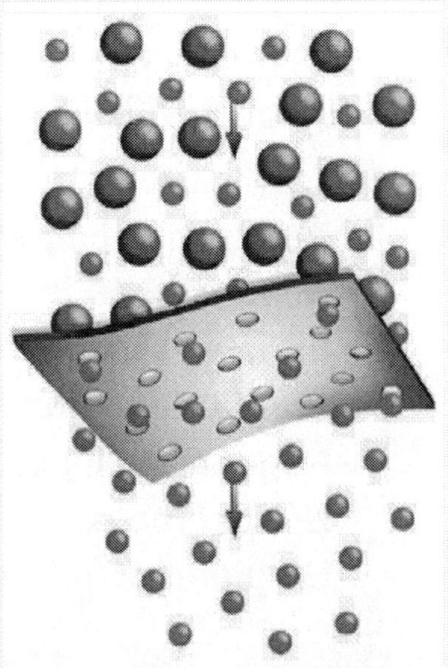

17 AIChE Spring Meeting, San Antonio, TX

16

ch2m. Axens IFP Group Technologies

Molecular Sieve Impurity Application

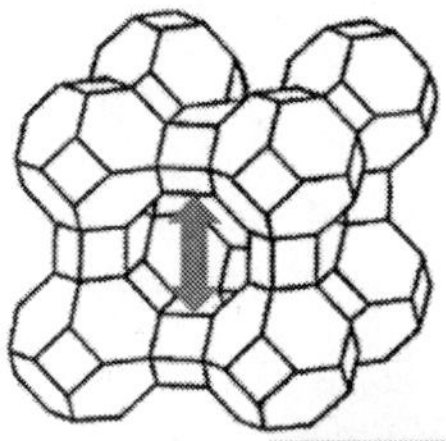

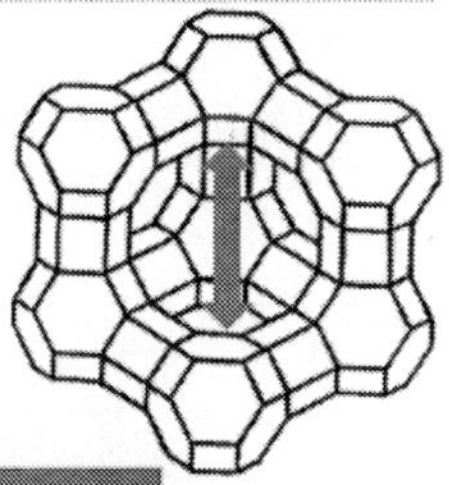

Increasing window size →

	3A	4A	5A	13X
Cation	K	Na	Ca	Na
Opening	3Å	4Å	5Å	~10Å
Application	Selective drying	Standard drying	Purification	Purification
Impurities	H_2O, NH_3	+ Methanol CO_2, H_2S, N_2	+ C1-SH C2-SH	+ all polar compounds

→ Increasing co-adsorption impurities/feed

Molecular Sieve Structure and Properties

- Zeolite molecular sieves belong to the class of aluminosilicates
- They are crystalline and highly porous materials
- The crystals have pores with a precisely defined diameter
- The most common commercial adsorbents zeolites are A and X

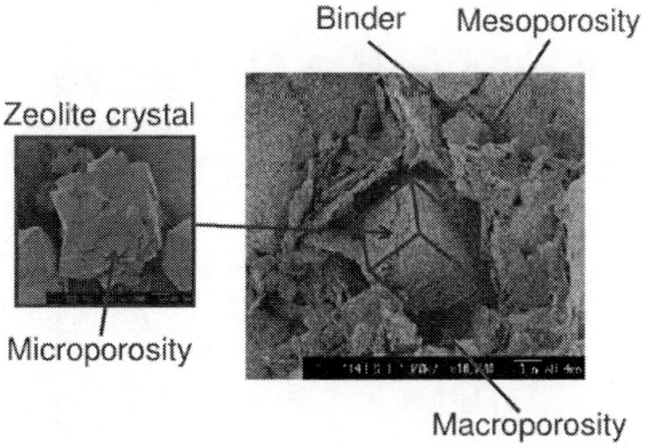

Zeolite A

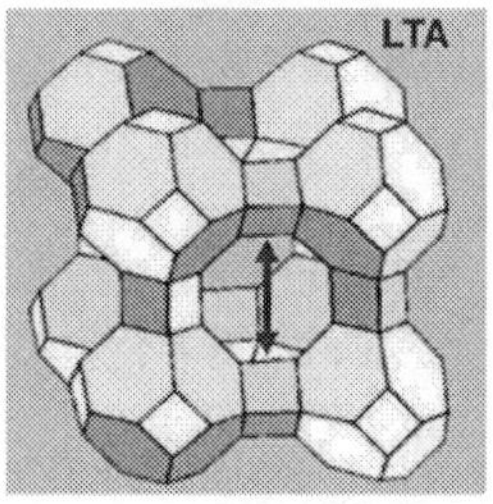

Zeolite X

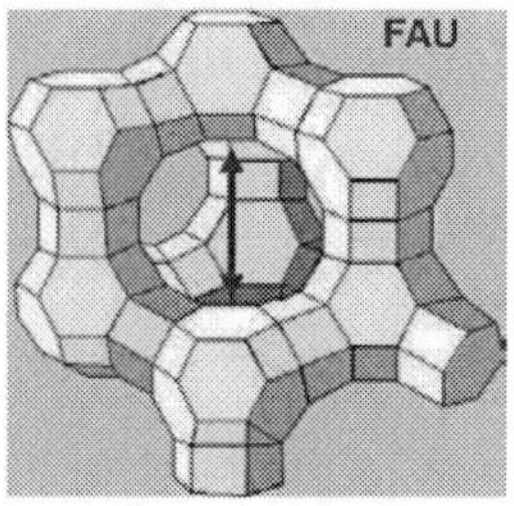

Fixed Bed Adsorption – A Dynamic Process

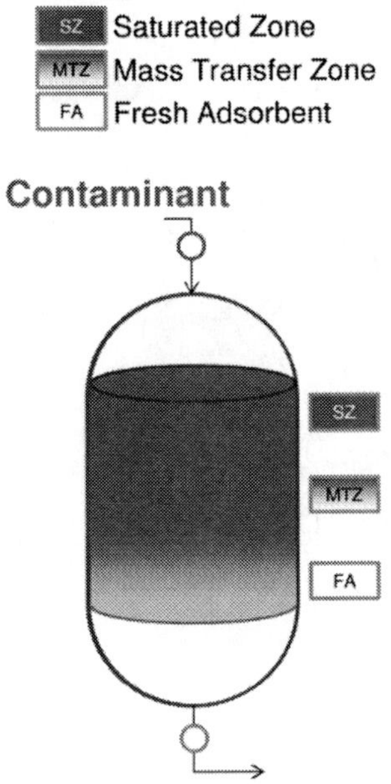

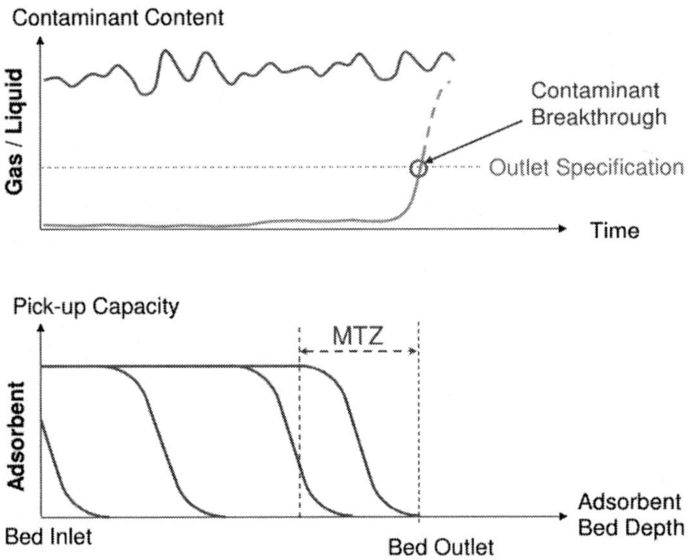

17 AIChE Spring Meeting, San Antonio, TX

ch2m: Axens IFP Group Technologies

Dynamic Adsorption and Design

In order to shorten the Mass Transfer Zone (MTZ):

- **Single layer, large sized molecular sieve**
  - **Low pressure drop ΔP**
  - **"Low" adsorption capacity / cycle length**
- **Double layer, large + small size molecular sieve**
  - **Slightly Higher pressure drop ΔP**
  - **High adsorption capacity / cycle length**

17 AIChE Spring Meeting, San Antonio, TX

ch2m: Axens IFP Group Technologies

Regeneration: Downflow or Upflow?

Regeneration Downflow

Adsorption

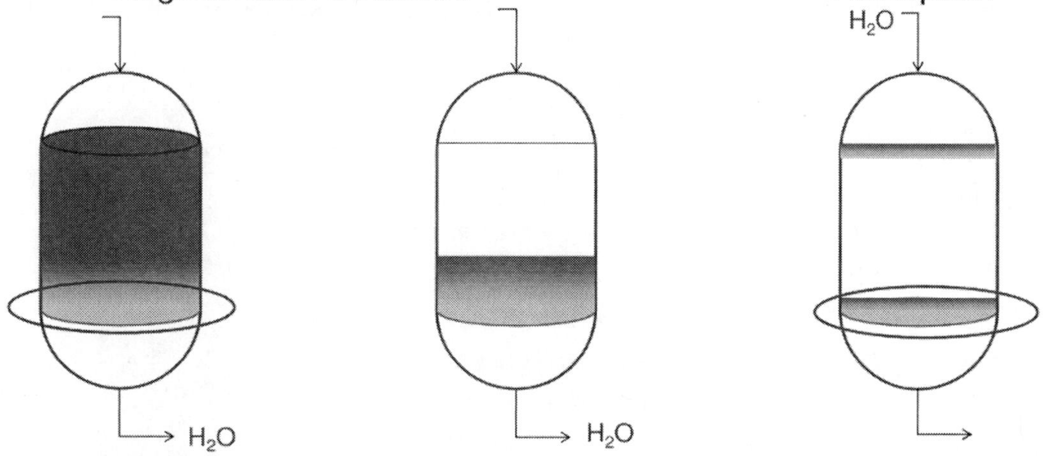

At the beginning of the regeneration phase, only few water molecules located at the bottom molecule are released

With time, the last water molecules are located in the bottom. The top layers are totally regenerated

When adsorption starts, there may have some water molecules at the bottom.
Potential earlier breakthrough

17 AIChE Spring Meeting, San Antonio, TX

21

ch2m: Axens IFP Group Technologies

Regeneration: Dowflow or Upflow?

Regeneration Upflow

Adsorption

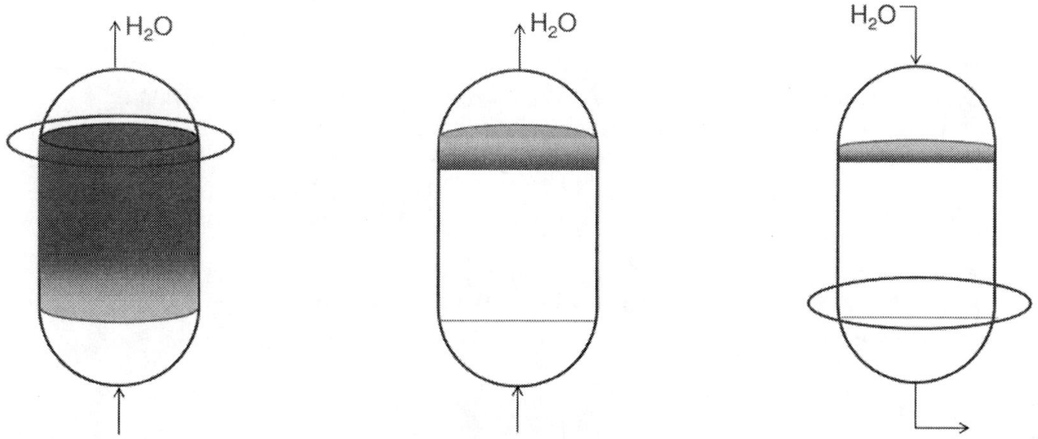

At the beginning of the regeneration phase, many water molecules located at the top molecule are released.

With time, the last water molecules are located in the top. The bottom layers are totally regenerated

When adsorption starts, there is not water molecules at the bottom.
No potential breakthrough

17 AIChE Spring Meeting, San Antonio, TX

22

ch2m: Axens IFP Group Technologies

Regeneration Gas Considerations

- **The following should be considered for proper regeneration:**

1. Source and quality **of regen gas:**
   - **Mercaptans, methanol, amines, or CO_2 will react at high temperature by providing side reactions on the surface of the molecular sieve**
   - **Coke formation or polymerization block the cages and limits therefore the capacity of the water molecules to access to the adsorption sites**
2. Pressure **of the available regen gas**
3. Quantity **of the available regen gas:**
   - **Reflux (retro condensation of water during the heating step)**
   - **Pressure drop and channeling**
4. Treatment/disposal **of regen gas**
5. Frequency **of regeneration**

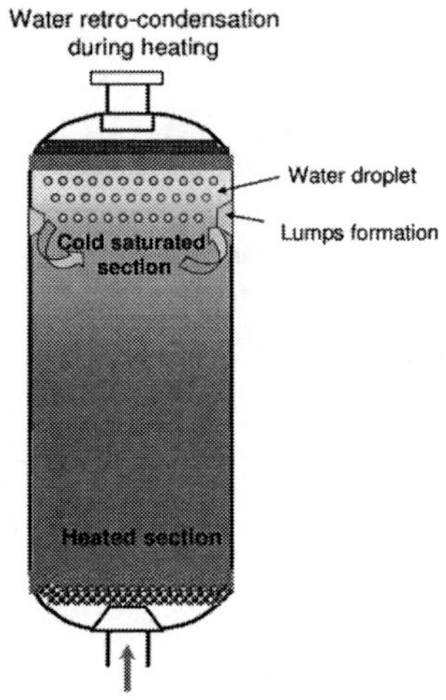

Challenges in Regeneration Considerations Source of Regen. Gas & Its Treatment

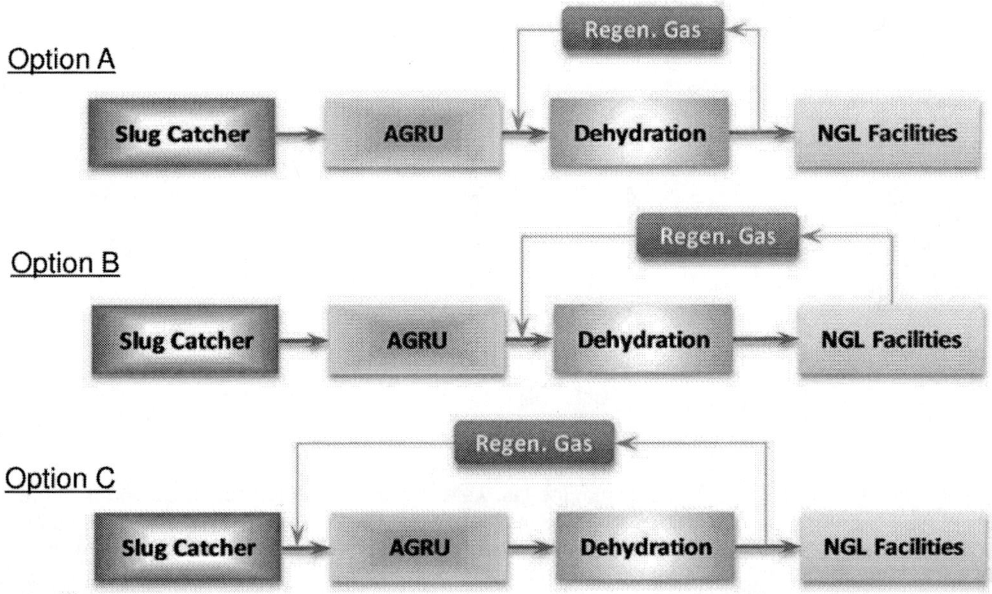

How Do You Operate Your Unit?

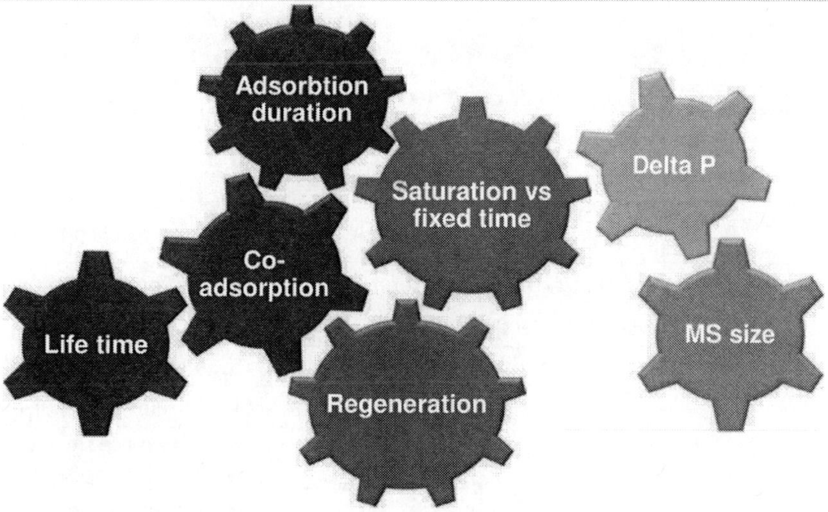

Many factors influence performance!

17 AIChE Spring Meeting, San Antonio, TX

25

ch2m: Axens IFP Group Technologies

Purpose of Optimization

Key Optimization Points	Design & Operating Considerations	Details
Treated gas specifications	Water, H2S, RSH, COS, CO2, and hydrocarbon removal	
	Pipeline specifications	Meet the water and hydrocarbon dew point specifications
	NGL recovery & LNG plants specifications	Meet the H2O content of 0.01 ppmv at treated gas stream
Optimize electricity consumption		
Minimize utility consumption		
Optimize CAPEX/OPEX		

17 AIChE Spring Meeting, San Antonio, TX

26

ch2m: Axens IFP Group Technologies

Process Efficiency

Key Optimization Points	Design & Operating Considerations	Details
Define process configuration alternatives	Single beds or combined beds	One kind of desiccant per bed or different kind of desiccant layers
	Number of beds	
	Feed pre-chilling requirement	
Optimize electricity consumption	Regeneration gas compressor	Minimize total compression energy
Maximize heat integration	Internal/external insulation	Reduce regeneration heat required
	Material selection	
Minimize utility consumption	Desiccant consumption	Optimize the number of bed and size
	Regeneration heater	Minimize fuel gas consumption
	Feed pre-chilling exchanger, optional	Optimize mole sieve dehydration package CAPEX/OPEX
		Reduce refrigerant consumption
	Regeneration gas cooler	Minimize cooling water/ electricity consumption

17AIChE Spring Meeting, San Antonio, TX

27

ch2m. Axens IFP Group Technologies

Adsorption System Design Specifications

Key Optimization Points	Design & Operating Considerations	Details
Number and size of trains	Determine operating envelope, range of capacity, and compositions	
	Determine most likely feed composition and rate	
Operating pressure	Inlet feed separator	
	Appropriate pressure drop/equipment	Inlet feed separator, adsorption beds, treated gas filter
Operating temperature	Inlet feed gas	Feed gas pre-chilling
	Regeneration temperature	
	Check compressors suction temperature	
	Check compressors discharge temperature	
	Check compressor coolers outlet temperature	
	Consider proper ambient temperature	
	Consider proper approach temperature	

17AIChE Spring Meeting, San Antonio, TX

28

ch2m. Axens IFP Group Technologies

Adsorption System Design Specifications (cont'd)

Key Optimization Points	Design & Operating Considerations	Details
Desiccant quality specifications	Type of desiccant	
	3A molecular sieve	
	4A molecular sieve	
	5A molecular sieve	
	13X molecular sieve	
	Combined beds	H_2S, mercaptans and COS removal
	Alumina	
	Regeneration cycle optimization	Number of adsorber vessels
		Regeneration temperature
		Regeneration flow rate
		Regeneration gas source
		Pressure cycling

Operational Challenges

- Amine carryover

- Glycol carryover

- Feed condensation

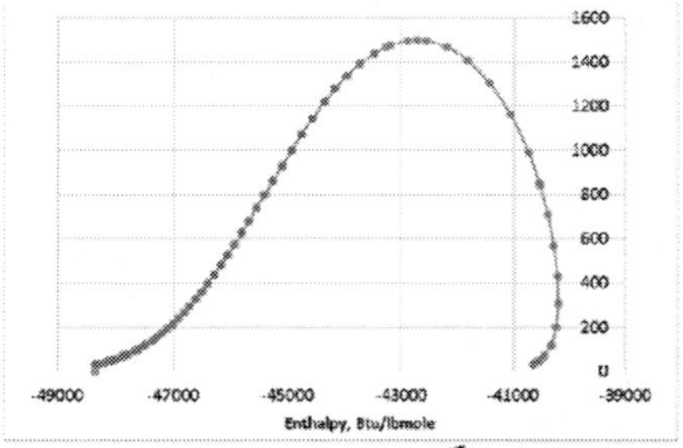

Impact of Upstream Processes

Foaming problems:
A potential for significant operational upsets and amine carryover to the driers

AGRU Feed Gas may contain:

1. N2
2. H2O
3. H2S
4. CO2
5. Hg
6. RSH
7. MeOH
8. MEG
9. Scale Inhibitors
10. Entrained HC Liquids

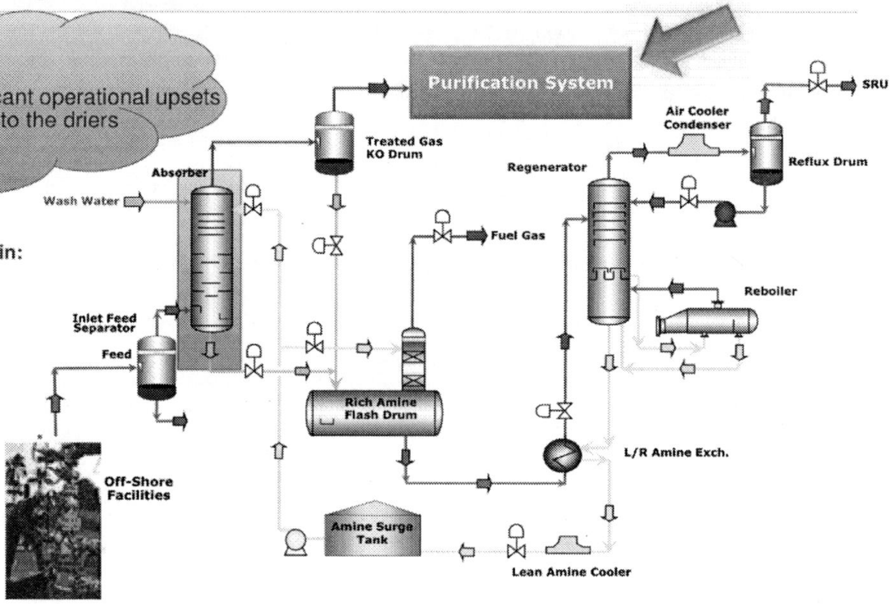

Table of Contents

Why is Drying Important?

- To prevent hydrate formation that can plug equipment

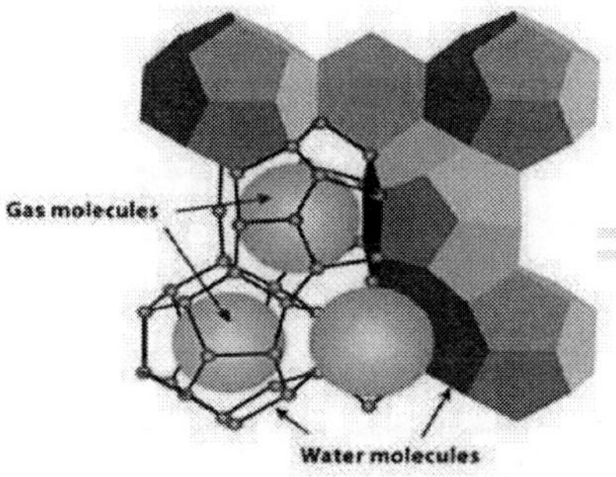

17AIChE
Spring Meeting, San Antonio, TX

33

ch2m: Axens IFP Group Technologies

Drying Product Options

- Drying can be accomplished with activated alumina or molecular sieve

Activated Alumina

Molecular Sieve

17AIChE
Spring Meeting, San Antonio, TX

34

ch2m: Axens IFP Group Technologies

Static Water Adsorption Capacity

- A molecular sieve's static water adsorption capacity is high at low pressures but it flat lines at higher pressures
- An activated alumina's static water adsorption capacity continually increases with the pressure and eventually exceeds a molecular sieve's capacity

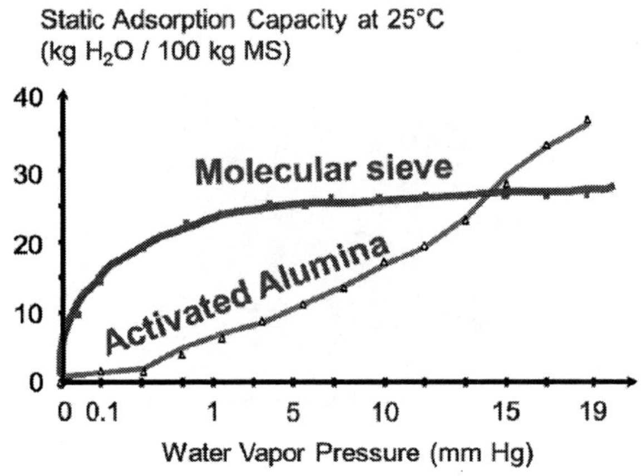

17AIChE Spring Meeting, San Antonio, TX

35

ch2m. Axens IFP Group Technologies

Dynamic Water Adsorption Capacity

- Loading both activated alumina and molecular sieve increases the dynamic water adsorption capacity of the vessel even after many regenerations

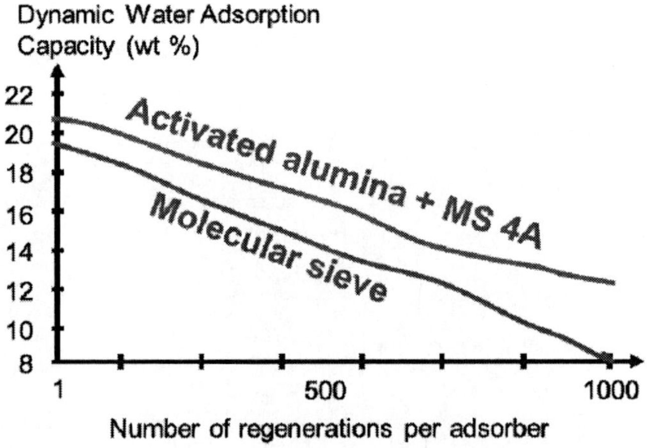

17AIChE Spring Meeting, San Antonio, TX

36

ch2m. Axens IFP Group Technologies

Drying Challenge and Solution

- Scenario: **Water specification is < 0.01 ppmv but there's a potential for intermittent free water in feed**
- Challenge: **A molecular sieve is required to meet spec but free water in the feed will quickly degrade it**
- Solution: **Load activated alumina ahead of the molecular sieve for protection from free water and increased dynamic water adsorption capacity**

17 AIChE Spring Meeting, San Antonio, TX

37

ch2m: Axens IFP Group Technologies

Process Flow Diagram

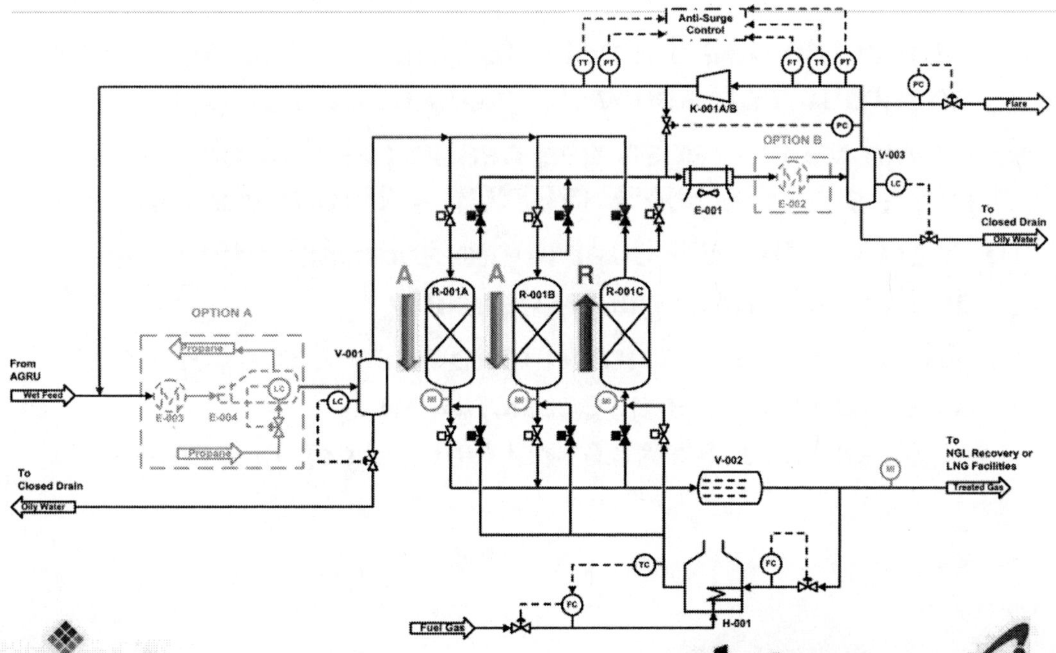

17 AIChE Spring Meeting, San Antonio, TX

38

ch2m: Axens IFP Group Technologies

Vendor Steps in System Sizing

1. Select the sieve based on feed composition and effluent requirements
2. Choose a step time, normally 8 - 16 hours
3. Choose (2 + 1) or (3 + 1) configuration
4. Calculate the cycles per life cycle period
5. Calculate the adsorbent requirements per bed
6. Adjust the adsorbent requirements based on life cycle period cyclic efficiency
7. Assume a bed diameter, calculate the MS depth, and check the adsorption pressure drop

17 AIChE Spring Meeting, San Antonio, TX

39

ch2m: Axens IFP Group Technologies

System Sizing (continued)

8. Adjust the bed diameter (6" increments) to get an adsorption pressure drop of about 5 - 6 psi
9. Calculate the regen gas heating and cooling requirements, 550°F (288°C), ~ 2 hour standby
10. Increase the MS quantity to accommodate the additional regen gas water duty
11. Check the regen flowrate to assure proper flow distribution during regeneration
    - Should be at least 0.01 psi/foot bed depth, bottom

17 AIChE Spring Meeting, San Antonio, TX

40

ch2m: Axens IFP Group Technologies

Why Not Contact Licensor for the Optimum Solution?

Description	Unit	Bidder A	Bidder B	Bidder C	Bidder D
No of Beds		3+1	3+1	3+1	4+1
Bed Diameter	mm	4180	3500	3200	3810
Bed Height	mm	13600	4800	7315	10668
Adsorption Time	hr	18	8	8	9
Heating Time	hr	3.5	1.5	1.4	1.9
Cooling Time	hr	2	0.5	0.8	0.8
Stand by Time	hr	0.5	0.65	0.45	0.8
Regeneration Flowrate	MMSCFD	104	96	53.1	59

17 AIChE Spring Meeting, San Antonio, TX

41

ch2m: Axens IFP Group Technologies

Table of Contents

- Basics of Adsorption
- Drying Challenges and Solutions
- **Mercury Removal Challenges and Solutions**
- Sulfur Compound Removal Challenges and Solutions
- Conclusions

17 AIChE Spring Meeting, San Antonio, TX

42

ch2m: Axens IFP Group Technologies

Why is Mercury Removal Important?

- To prevent equipment/piping corrosion and failures
- To prevent health and safety issues
- To prevent downstream catalyst contamination
  (i.e. catalysts used in petrochemistry)

Hg^0

Mercury Occurrence and Problems

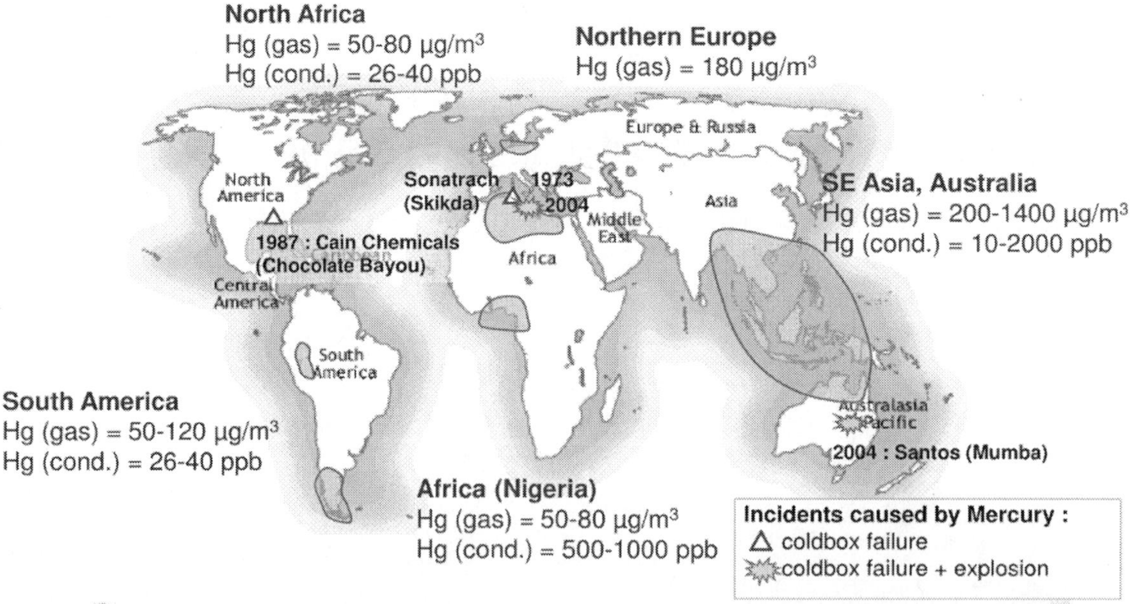

Mercury Impact: Health Safety

- **Exposure Levels***
  - <50 μg Hg /m^3
    TWA (Time Weighted Average) for 8 hours work exposure
  - >10 mg /m^3
    NIOSH Immediately Dangerous To Life or Health (IDLH)

*OSHA (Occupational Safety & Health Administration) regulations (www.osha.gov)

Mercury Impact: Catastrophic Damage

- **Skikda Explosion in Algeria in 1973**

$$Al + Hg^0 \rightarrow HgAl$$

$$HgAl + 3H_2O \rightarrow Al_2O_3 + 3H_2 + Hg^0$$

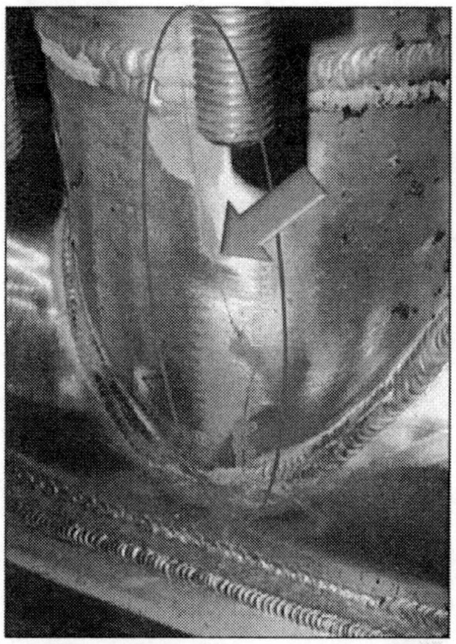

Mercury Embrittlement in LNG/NGL Facilities

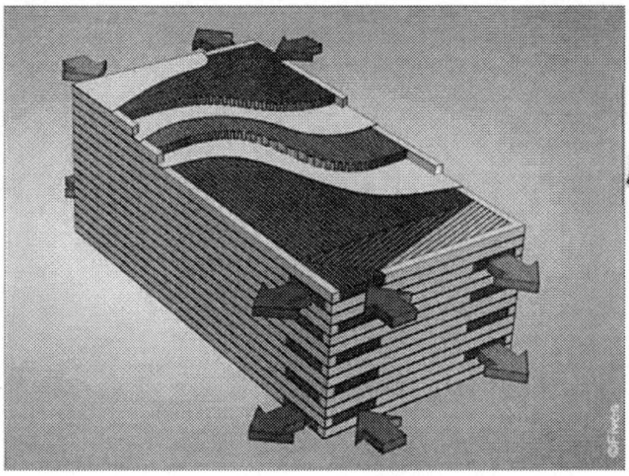

Brazed Aluminum Plate Fin Heat Exchanger (BAHX)

Mercury Removal History

- In late 1970's, the first mercury removal adsorbent for natural gas was developed

- Trapping mechanism is chemisorption
- Metallic mercury (elemental Hg^0) is trapped forming red crystalline HgS (cinnabar)

- Cinnabar formation (a <u>fast</u> reaction):

$$Hg^0 + 2\,CuS \rightarrow HgS + Cu_2S$$

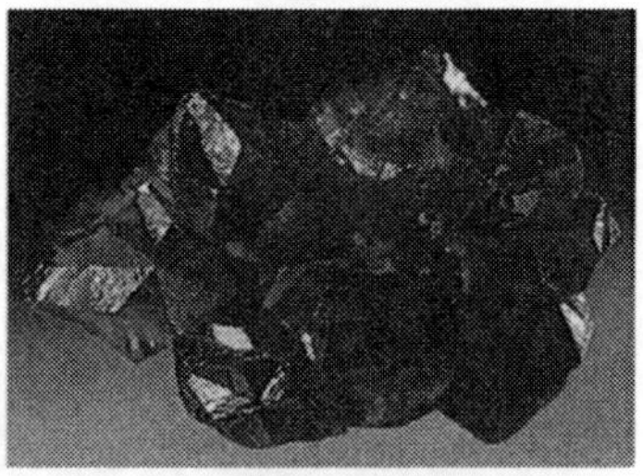

Cinnabar

Mercury Adsorption Reaction

- **Mercury is removed by reacting it with copper sulfide on an adsorbent to irreversibly create cinnabar (HgS)**

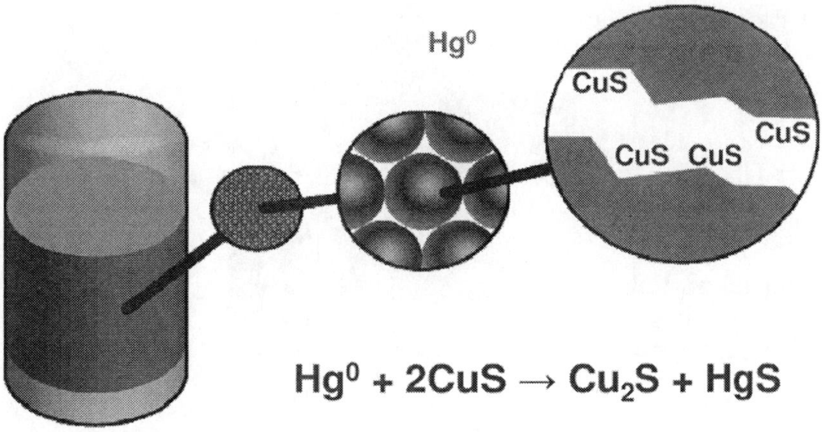

$$Hg^0 + 2CuS \rightarrow Cu_2S + HgS$$

17 AIChE Spring Meeting, San Antonio, TX

49

ch2m: Axens IFP Group Technologies

Mercury Adsorption Reaction

- **Mercury is removed by reacting it with copper sulfide on an adsorbent to irreversibly create cinnabar (HgS)**

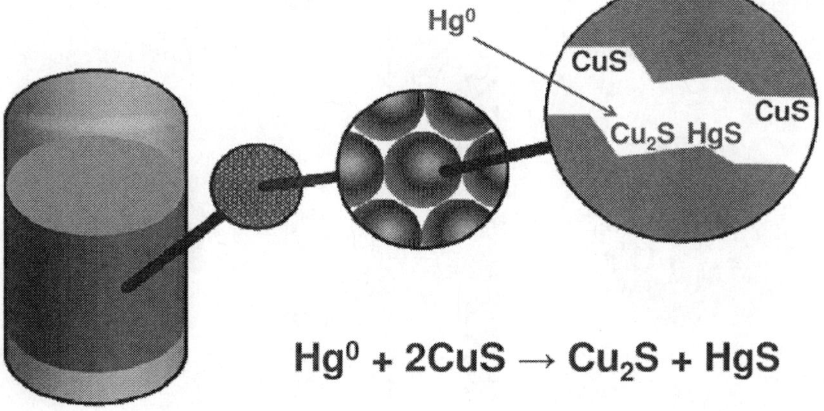

$$Hg^0 + 2CuS \rightarrow Cu_2S + HgS$$

17 AIChE Spring Meeting, San Antonio, TX

50

ch2m: Axens IFP Group Technologies

Mercury Adsorption Mechanism Improvements

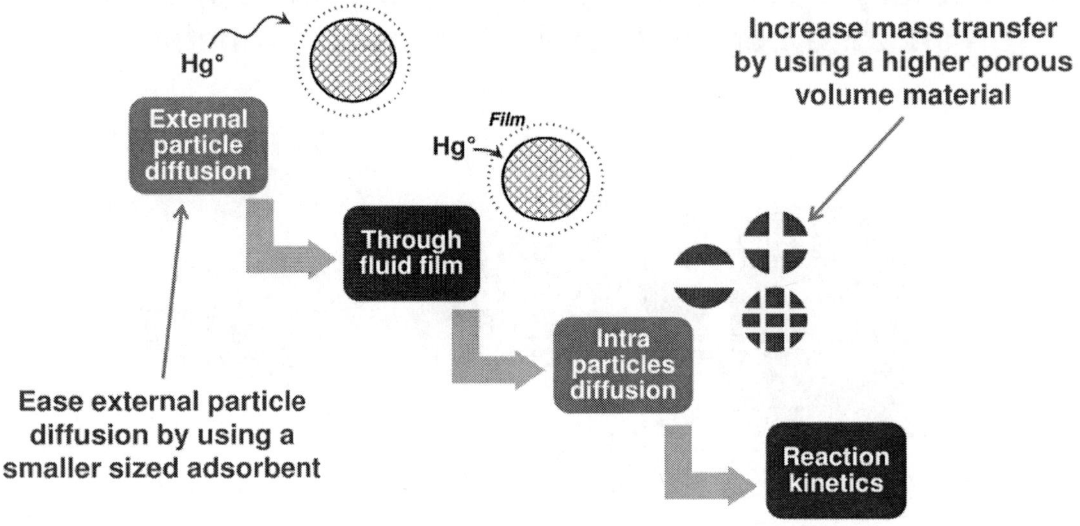

17AIChE Spring Meeting, San Antonio, TX

51

ch2m Axens IFP Group Technologies

Optimized Adsorbent Utilization

- Adsorbent with larger particles and low macroporosity is ineffective

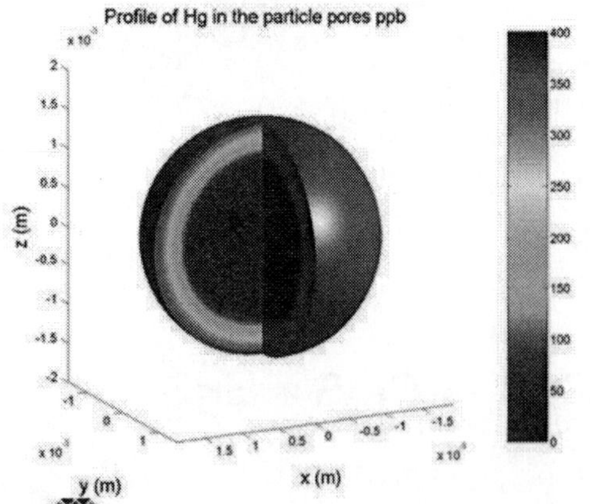

- Adsorbent with smaller particles and high macroporosity is fully utilized

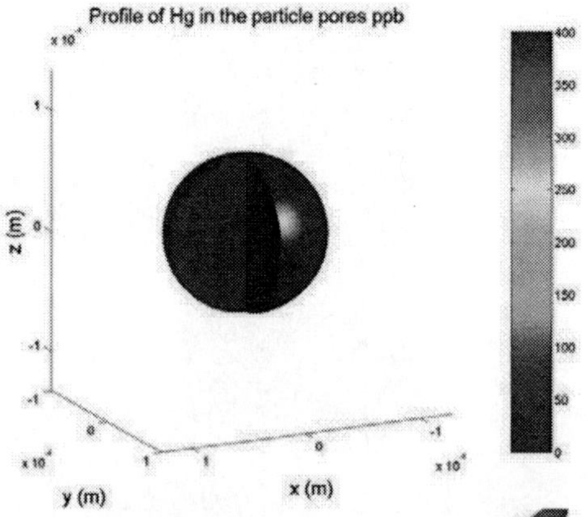

17AIChE Spring Meeting, San Antonio, TX

52

ch2m Axens IFP Group Technologies

Mercury Removal Challenge and Solution

- Scenario: **Mercury specification is <0.01 µg/Nm^3**
- Challenge: **The bulk copper sulfide adsorbent had premature mercury breakthrough and cementing problems after only two weeks**

Cemented bulk copper sulfide adsorbent

- Solution: **Use a porous alumina based copper sulfide adsorbent to improve mechanical resistance and mass transfer**

17AIChE Spring Meeting, San Antonio, TX

53

ch2m: Axens IFP Group Technologies

Product Comparison

- **Bulk copper sulfide adsorbent**
  - x Poor mechanical resistance
  - x Poor mass transfer properties with no pores and insufficient contact time
  - x Had cementing and premature breakthrough

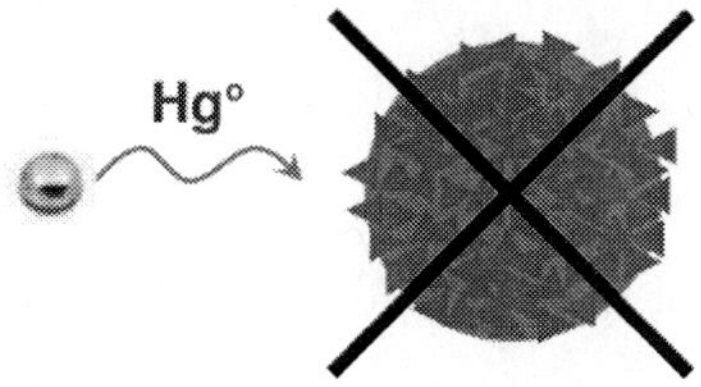

- **Optimized, porous copper sulfide adsorbent**
  - ✓ Good mechanical resistance
  - ✓ Good mass transfer properties with small adsorbent particles and large pores
  - ✓ Had no pressure drop issues or premature breakthrough

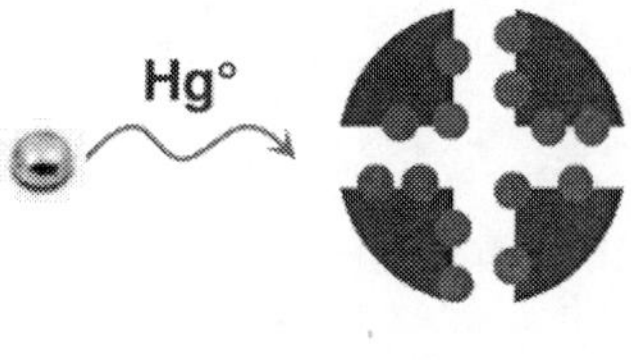

17AIChE Spring Meeting, San Antonio, TX

54

ch2m: Axens IFP Group Technologies

LNG/NGL Mercury Removal Processing Scheme

- **In the 2000's, the location of mercury removal in the natural gas process chain was moved upstream, which required a much more robust adsorbent**

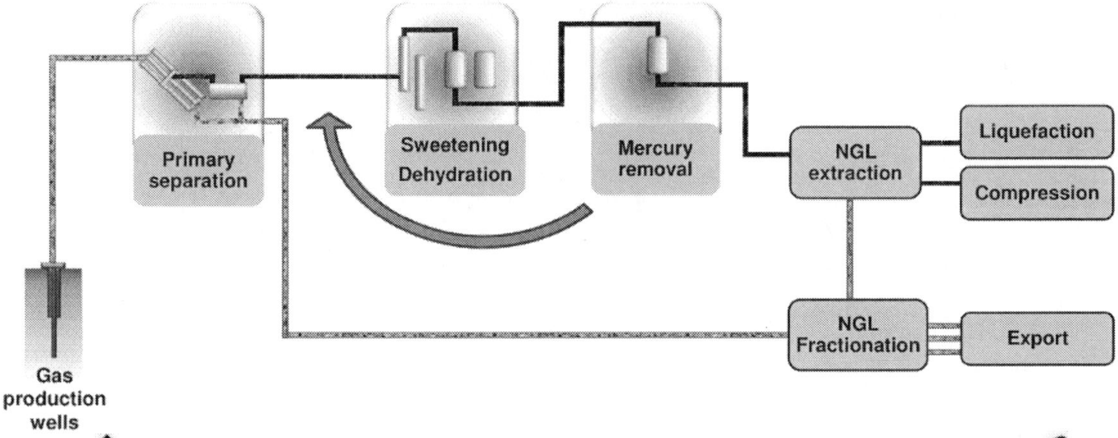

Upstream Positioned MRU

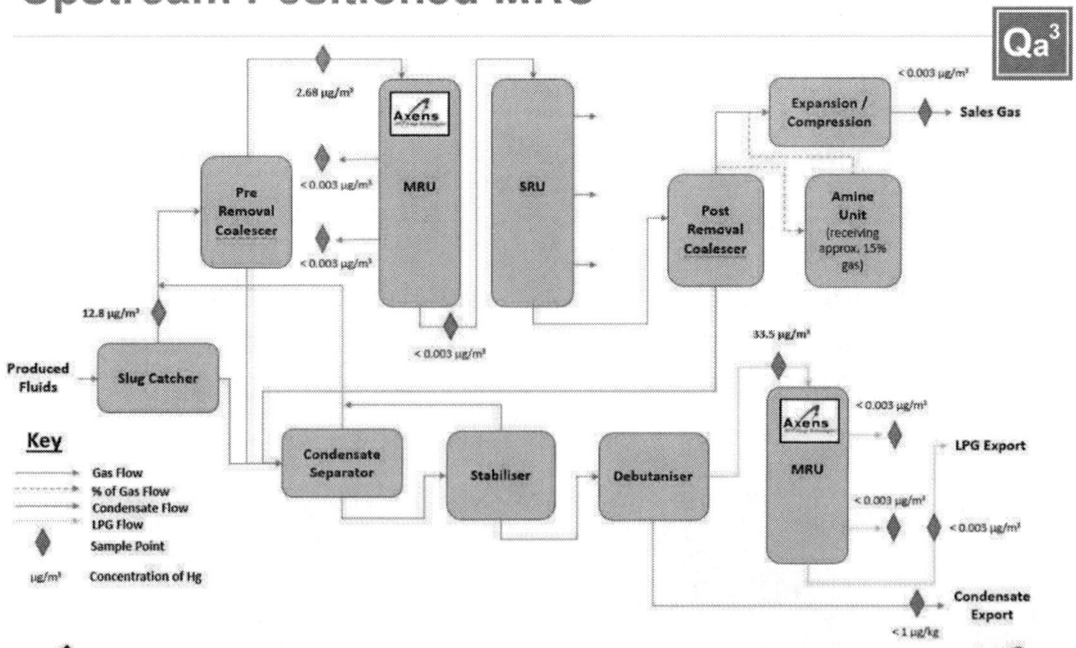

LNG/NGL Mercury Removal Processing Scheme

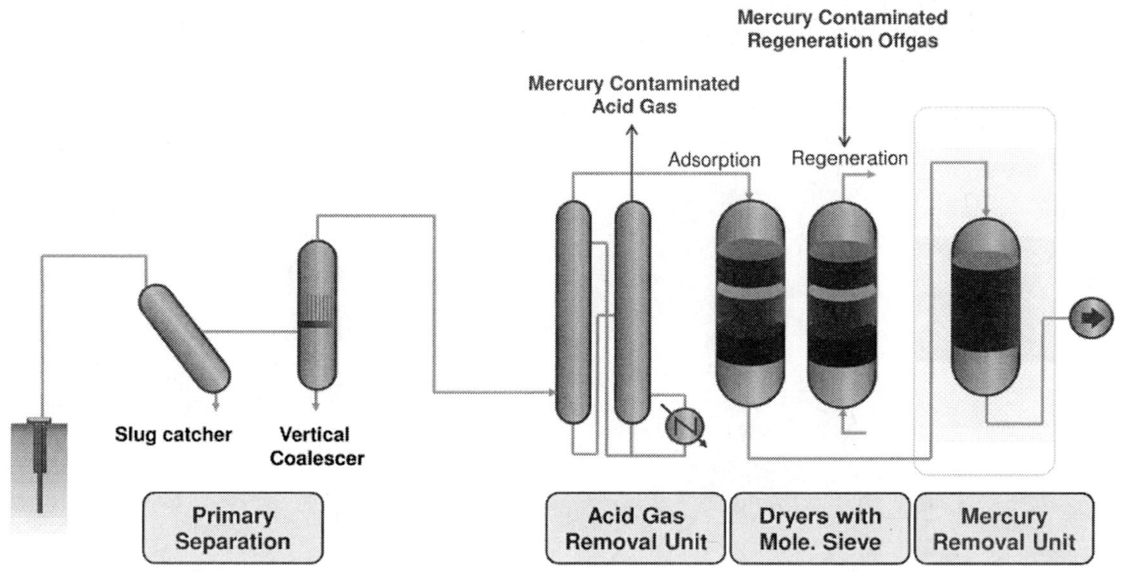

LNG/NGL Mercury Removal Processing Scheme

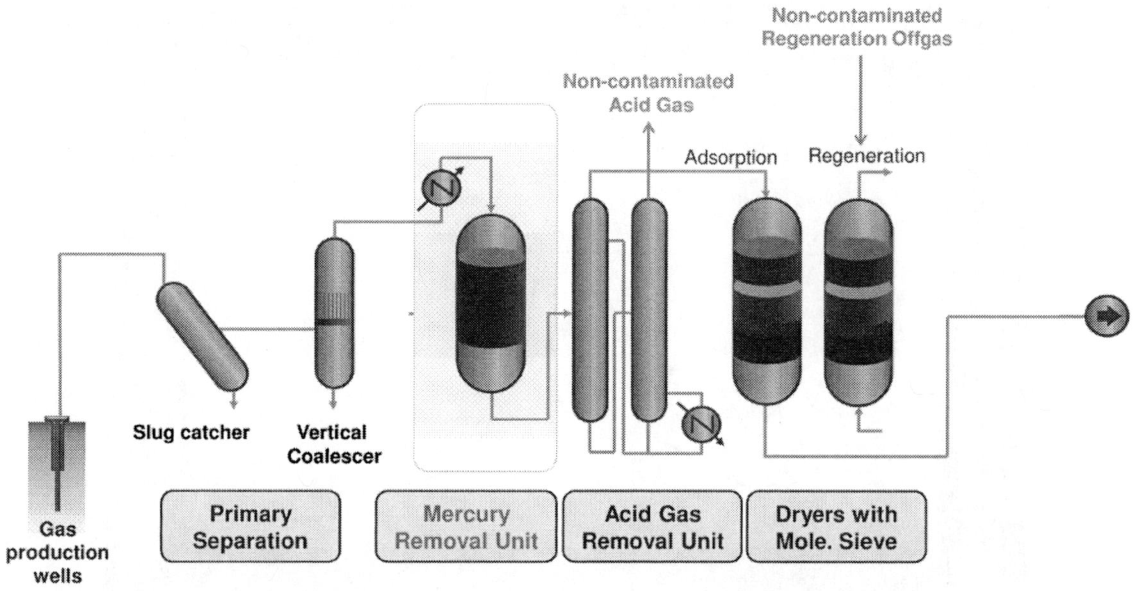

Where to Measure Mercury within the Process ?

Qa3

Dependent on presence and position of Mercury Removal Units (MRU)

- MRU placed after acid gas removal and dehydration

MRU Downstream in the Process

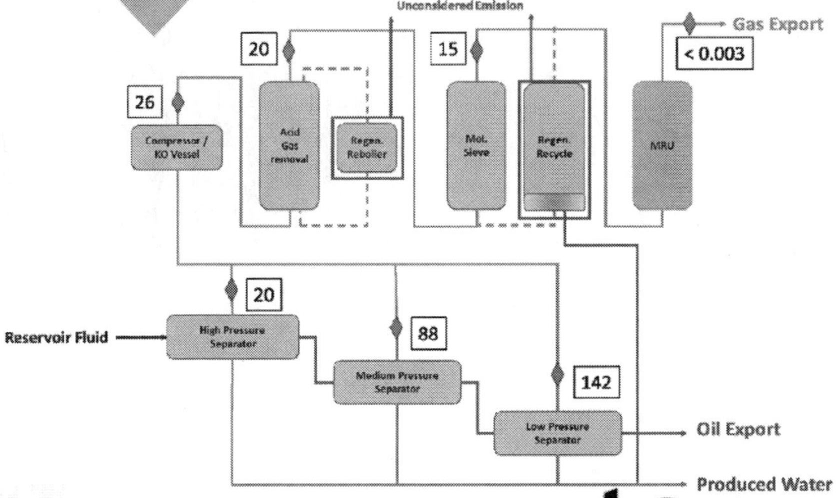

17 AIChE Spring Meeting, San Antonio, TX

59

ch2m. Axens IFP Group Technologies

Mercury Trend Data

Qa3

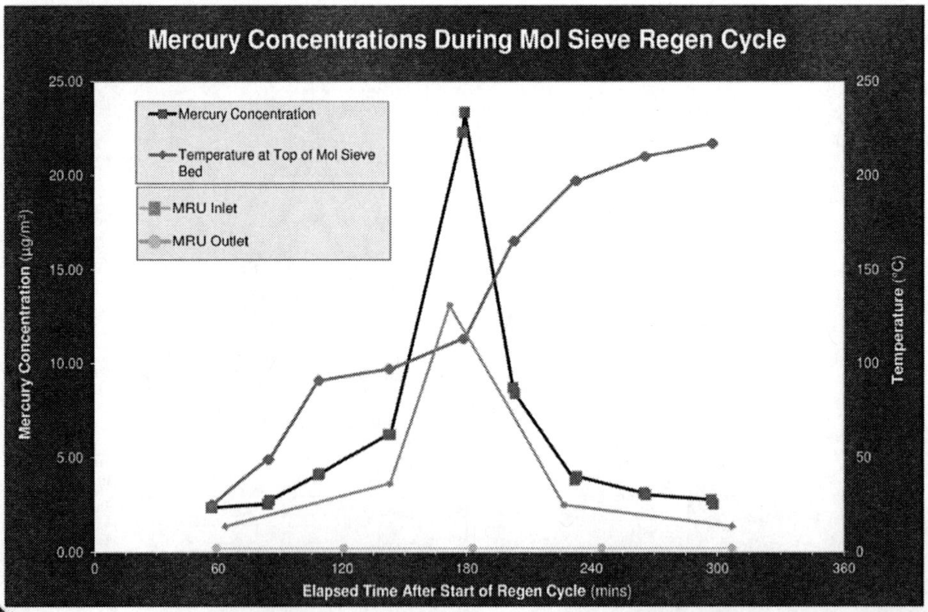

17 AIChE Spring Meeting, San Antonio, TX

60

ch2m. Axens IFP Group Technologies

Mercury Removal Challenge and Solution

- Scenario: A natural gas plant is trying to decide where to locate a new mercury removal unit (MRU)
- Challenge: Putting the MRU downstream of the sweetening dehydration unit allows the off-gases and upstream equipment to be contaminated with mercury
- Solution: Put the MRU upstream of the sweetening dehydration and use the latest advances in adsorbent technology to have successful performance

17 AIChE Spring Meeting, San Antonio, TX   61   ch2m: Axens IFP Group Technologies

Consequences of MRU Relocation

- Higher mercury inlet concentration
- Higher operating pressure
  - Increased density and viscosity
  - Decreased mercury diffusion coefficient
- Wet acid gases (CO_2 and H_2S) with 100% relative humidity
- Liquids carry-over (H_2O and hydrocarbons)
  - Leaching of active phase and adsorbent fines formation/cementation
  - Capillary condensation of H_2O and C_5+
- Therefore, need very robust mercury removal solutions:
  - Suitable adsorbents with stronger mechanical properties, higher capacity, higher accessibility to a stable active phase, and no capillary condensation
  - Adapted powerful modeling tools for vessel design and prediction of performance

17 AIChE Spring Meeting, San Antonio, TX   62   ch2m: Axens IFP Group Technologies

Challenges in Mercury Removal from Hydrocarbon Condensate

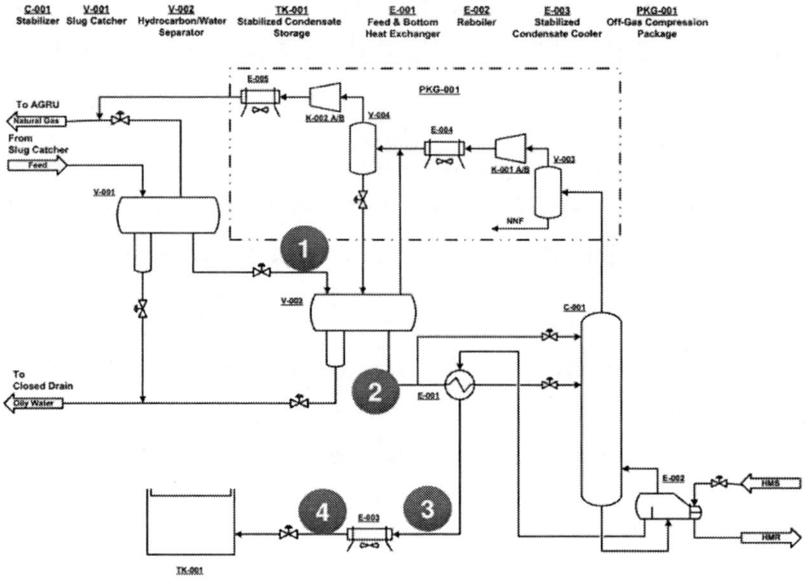

17 AIChE Spring Meeting, San Antonio, TX

63

ch2m. Axens IFP Group Technologies

Portable Pilot to Test Mercury Concentration

- Mercury inlet concentrations can be intermittent
- A portable pilot device can track mercury concentration over time (3-6 months) by taking a slip stream from the feed
- After the test, the adsorbent can be analyzed to determine the amount of mercury in the feed and the adsorption profile in order to complete a feasibility study

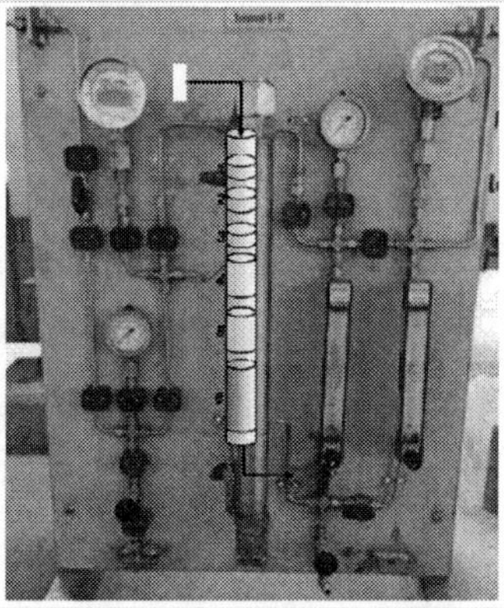

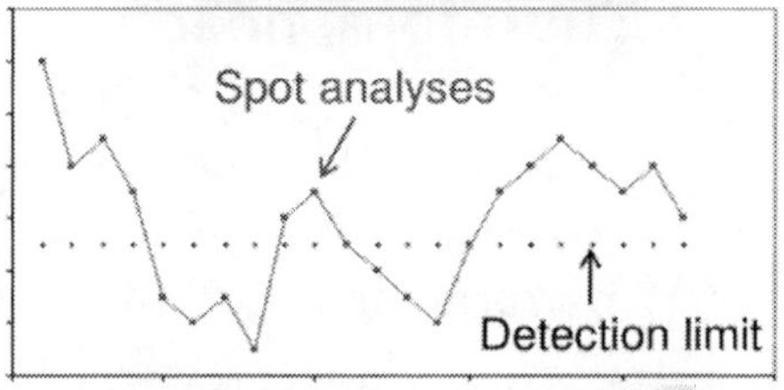

17 AIChE Spring Meeting, San Antonio, TX

64

ch2m. Axens IFP Group Technologies

Table of Contents

17 AIChE Spring Meeting, San Antonio, TX

ch2m: Axens IFP Group Technologies

Why is Sulfur Removal Important?

- To meet product specification
- To prevent corrosion
- To minimize sulfur emissions and HSE incidents

17 AIChE Spring Meeting, San Antonio, TX

ch2m: Axens IFP Group Technologies

Sources of H_2S & Organic Sulfur

- Sour primary natural gas production **(market depends upon mass of H_2S daily removal and access to gas gathering system and larger S removal systems)**
- Associated gas **produced with "sour" crude oil (market dependent upon price of crude oil)**
- Liquid Hydrocarbon Treatment – Copper Strip **(growing market)**
- Geothermal gas **(mainly for local air quality and CO_2 utilization, growing market)**
- Biogas – Landfill and WWTP biogas **(market dependent upon incentives and local air quality issues)**
- Storage Tanks and Transfer Points **(market dependent upon price of crude and local air quality issues)**

Gas Rate & H_2S Quantity (total daily mass) are primary criteria for selection of most economical technology

17 AIChE Spring Meeting, San Antonio, TX

67

ch2m: Axens IFP Group Technologies

H_2S & Organic Sulfur Removal Alternatives

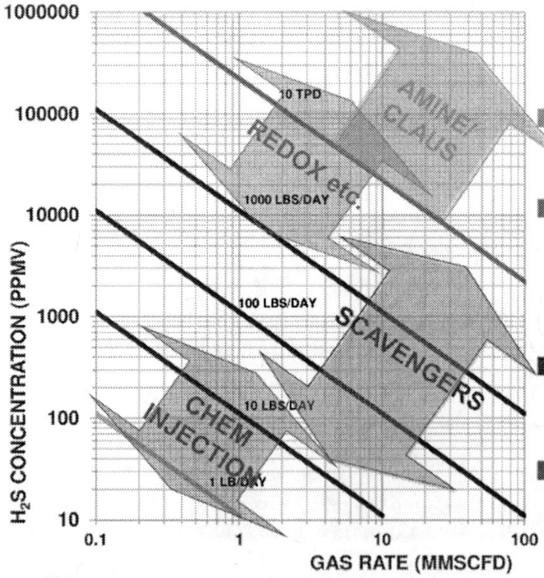

CAPEX	OPEX
Multi MM$	Few cts/lb
Few MM$	~$1/lb
>$50k	<$5/lb
<<$50k	>$10/lb

17 AIChE Spring Meeting, San Antonio, TX

68

ch2m: Axens IFP Group Technologies

Sulfur handling

- A large variety of sulfur impurities can be found…

H_2S RSH COS CS_2

…and they need to be removed for product specification issues

- Suited Technical Solutions
  - Solvents for bulk removal
    - Chemical solvents (amines)
    - Physical solvents
  - Solid sorbents for polishing
    - Molecular sieves
    - Scavengers
    - Metal Oxides

17 AIChE Spring Meeting, San Antonio, TX

69

ch2m: Axens IFP Group Technologies

Pros & Cons of Solid Sorbents

Molecular Sieves	Scavengers	Metal Oxides
Regenerable	**Guard Bed**	**Guard Bed**
High affinity for polar molecules like H_2S or RSH	Can operate in water saturated streams	One single product can handle a wide range of sulfur species
Can handle large S inlet content	Fast reaction kinetics with H_2S	Large sulfur pick-up at low or high temperatures
More complex process scheme	Easy spent management	Most suited for handling low sulfur inlet levels
Regeneration gas treatment	Most suited for handling low sulfur inlet levels	
Competition (H_2O, CO_2)	Limited kinetics reaction with other sulfur species	
Limited affinity for COS and CS_2		

17 AIChE Spring Meeting, San Antonio, TX

70

ch2m: Axens IFP Group Technologies

Desirable Characteristics for Scavenger Media

- Applications include:
  - Odour control and Sulfur emission (SOx ...)
  - Methane valorization (associated gases, Landfill gases, Biogas)
  - CO_2 processing

- Best when formed as a granular solid with iron and other metal oxides on natural base
- Should be water tolerant and able to operate in presence of O_2 in the gas stream
- Should be non-hazardous and non-corrosive
- Should be easy and safe to remove spent adsorbent material especially with O_2 present Should be easy to use with potentially unattended operation
- When possible, spent products should be non-hazardous for disposal per local regulations (e.g. non-hazardous promoted)
- Favorable when spent material qualifies for landfill on location or use as road-mix material
- Generally a low capital investment compared with regenerable processes

17 AIChE Spring Meeting, San Antonio, TX

71

ch2m. Axens IFP Group Technologies

Solid Scavenger Advantages and Disadvantages

Parameter	Advantage	Disadvantage
Capital Cost	Low to moderate install cost	Requires simple containment vessel
Operating Cost	Low to moderate operating cost	Requires added clean out cost
Engineering	Easily constructed and installed	Requires a unique design for each specific application
Operation Ease	Easy/minimal operator attention	Requires periodic replacement
Turndown Operation	Flexible if conditions are below minimum design conditions	Treatment results can be erratic
Utilities	No utilities required generally	Some products require water injection
Process Upsets	No media foaming/adjustments	
Organic Sulfurs Besides H_2S	Some products remove small amounts of mercaptans and other organic sulfurs	Some products can not remove organic sulfurs or will convert organic sulfurs to disulfides

17 AIChE Spring Meeting, San Antonio, TX

72

ch2m. Axens IFP Group Technologies

Solid Scavenger Advantages and Disadvantages

Parameter	Advantage	Disadvantage
Supporting Equipment	Requires only bulk liquid removal upstream	
Intermittent Operation	Easy to start and stop; can be used as standby for larger sulfur removal systems	
Process Fluids Treated	Can treat gas or liquid streams based on product selection	
Final H_2S Level	Some products can treat to as low as 0.1 ppmv	Some products can only treat to 4 ppmv
Interferences	Depends on product selection	Crude oil, corrosion inhibitors, film barriers, etc.

A Focus on COS Handling

Traces of COS can be naturally found…

…but can also been generated along the way in the purification section

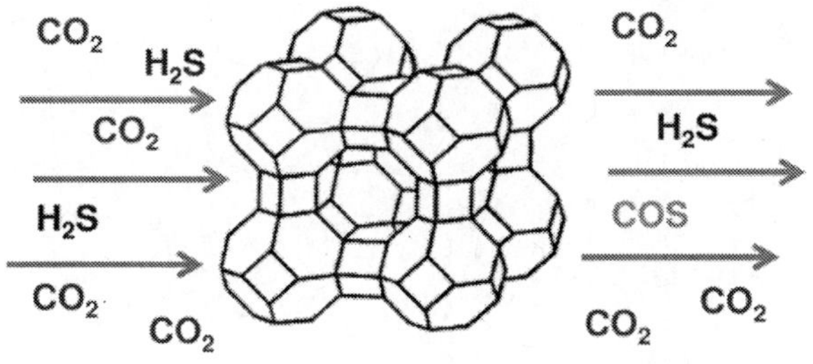

In addition, COS has a very limited affinity for Molecular Sieves

→ Downstream contamination

Solutions are needed to handle COS contamination

A Focus on COS Handling

Hydrolysis

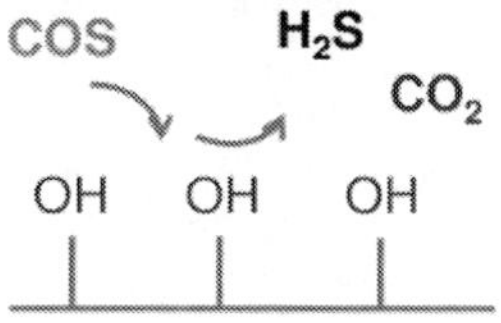

Dissociation of COS through reaction at the metal oxide (e.g. Al_2O_3) surface

A layer of alumina can both perform hydrolysis of COS and adsorption of H_2S & CO_2

Adsorption

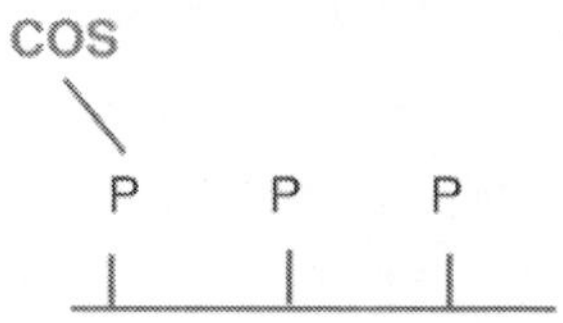

Promoted metal oxide (e.g. Al_2O_3) can retain COS.in a regenerative process

Prevention

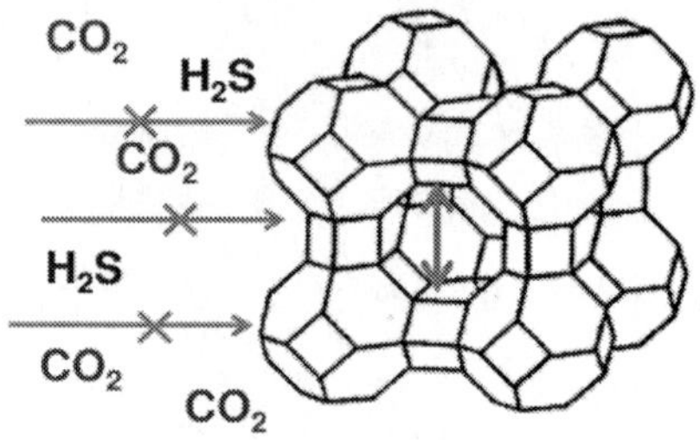

A carefully chosen MS pore size aperture (and/or binder formulation) will prevent CO_2 & H_2S reaction and hence prevent COS formation

17 AIChE Spring Meeting, San Antonio, TX

75

ch2m: Axens IFP Group Technologies

A Focus on COS Handling

Hydrolysis

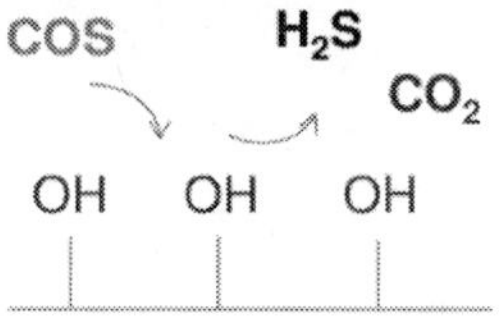

Dissociation of COS through reaction at the metal oxide (e.g. Al_2O_3) surface

A layer of alumina can both perform hydrolysis of COS and adsorption of H_2S & CO_2

Adsorption

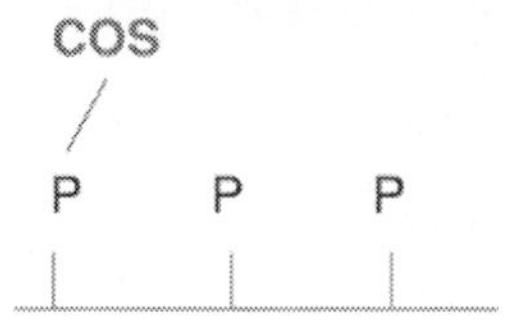

Promoted metal oxide (e.g. Al_2O_3) can retain COS in a regenerative process

Prevention

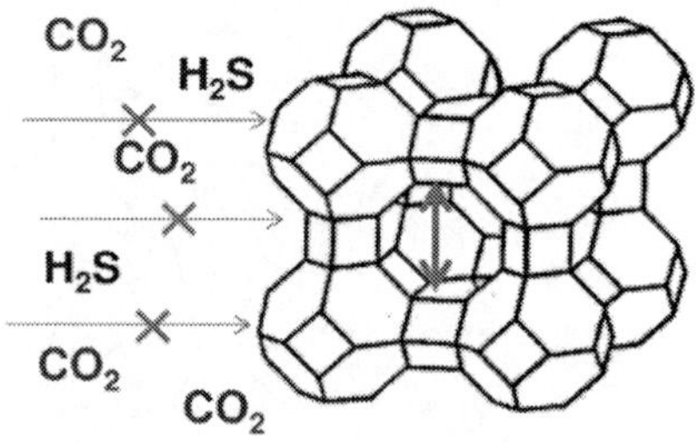

A carefully chosen MS pore size aperture (and/or binder formulation) will prevent CO_2 & H_2S reaction and hence prevent COS formation

17 AIChE Spring Meeting, San Antonio, TX

76

ch2m: Axens IFP Group Technologies

Sulfur Removal Challenge and Solution

- Scenario: **COS is being formed on a molecular sieve**
- Challenge: **The molecular sieve is not capturing the COS it creates**
- Solution: **Use hydrolysis or adsorption downstream to dissociate or adsorb COS**

17 AIChE Spring Meeting, San Antonio, TX

77

ch2m: Axens IFP Group Technologies

Table of Contents

- Basics of Adsorption
- Drying Challenges and Solutions
- Mercury Removal Challenges and Solutions
- Sulfur Compound Removal Challenges and Solutions
- **Conclusions**

17 AIChE Spring Meeting, San Antonio, TX

78

ch2m: Axens IFP Group Technologies

Conclusions

- Contaminant's management along the midtream process is challenging
- A clear view of impurities presence and their inlet concentration is needed at the project stage:
  - The design of the complete purification scheme can be impacted
  - The selection of the right adsorbent at the right location is crucial
- There is a large variety of adsorbent options available to solve many contamination problem or Troubleshooting operational issues

17 AIChE Spring Meeting, San Antonio, TX

79

ch2m: Axens IFP Group Technologies

Questions?

17 AIChE Spring Meeting, San Antonio, TX

80

ch2m: Axens IFP Group Technologies

Iron-Containing Carbon Molecular Sieve Membranes for Advanced Olefin/Paraffin Separations

Yu-Han Chu, William J. Koros
- Georgia Institute of Technology
David F. Yancey, Mark Brayden, Marcos Martinez, and Liren Xu
- The Dow Chemical Company

Georgia Institute of Technology   DOW   17 AIChE Spring Meeting, San Antonio, TX

2017 AIChE National Spring Meeting
"Advances in Gas Processing Design, Purification, and Separation" session
March 28th Tuesday 1:30-2:00PM.

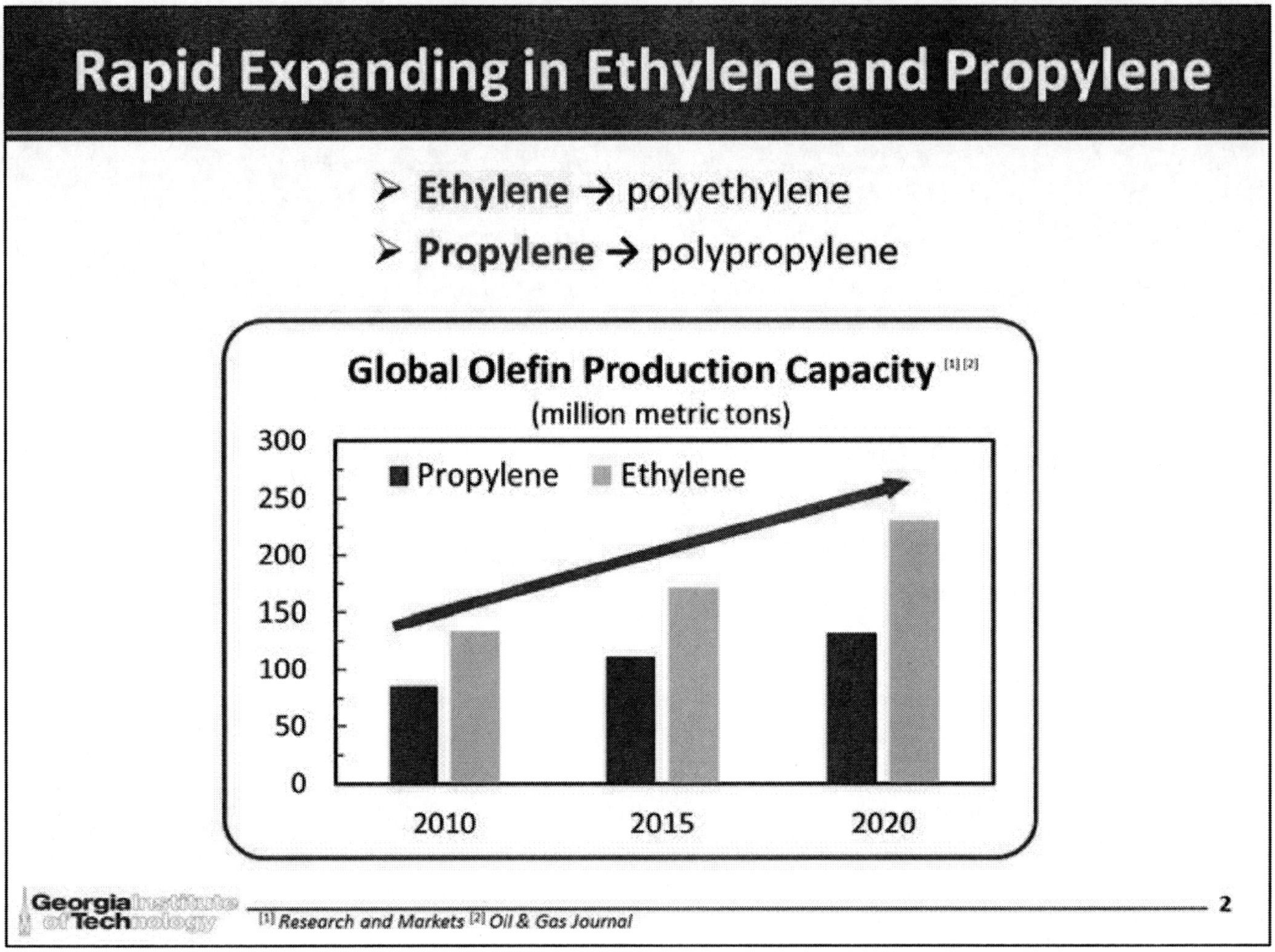

- Light olefins, such as ethylene and propylene, are important building blocks in chemical industry.
- The demands for ethylene and propylene are expanding at a rapid pace.

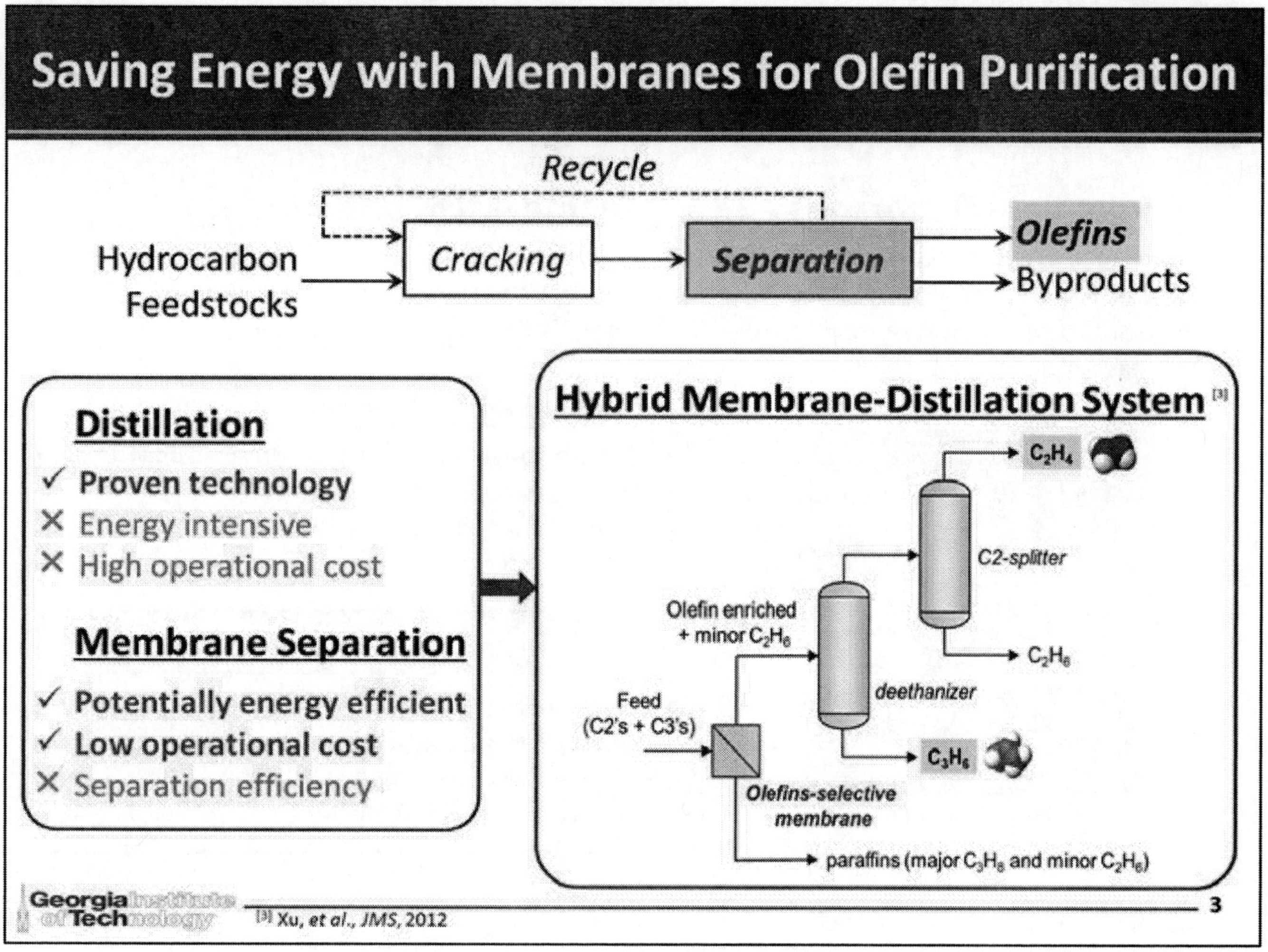

- A series of separation processes is required to recover valuable olefins from byproducts.
- Membrane technology may provide an attractive alternative to traditional thermally driven methods for gas separation.
- In this work, olefin-selective membranes are described for use in a hybrid membrane-distillation process to retrofit the current distillation process to lower the overall energy usage.

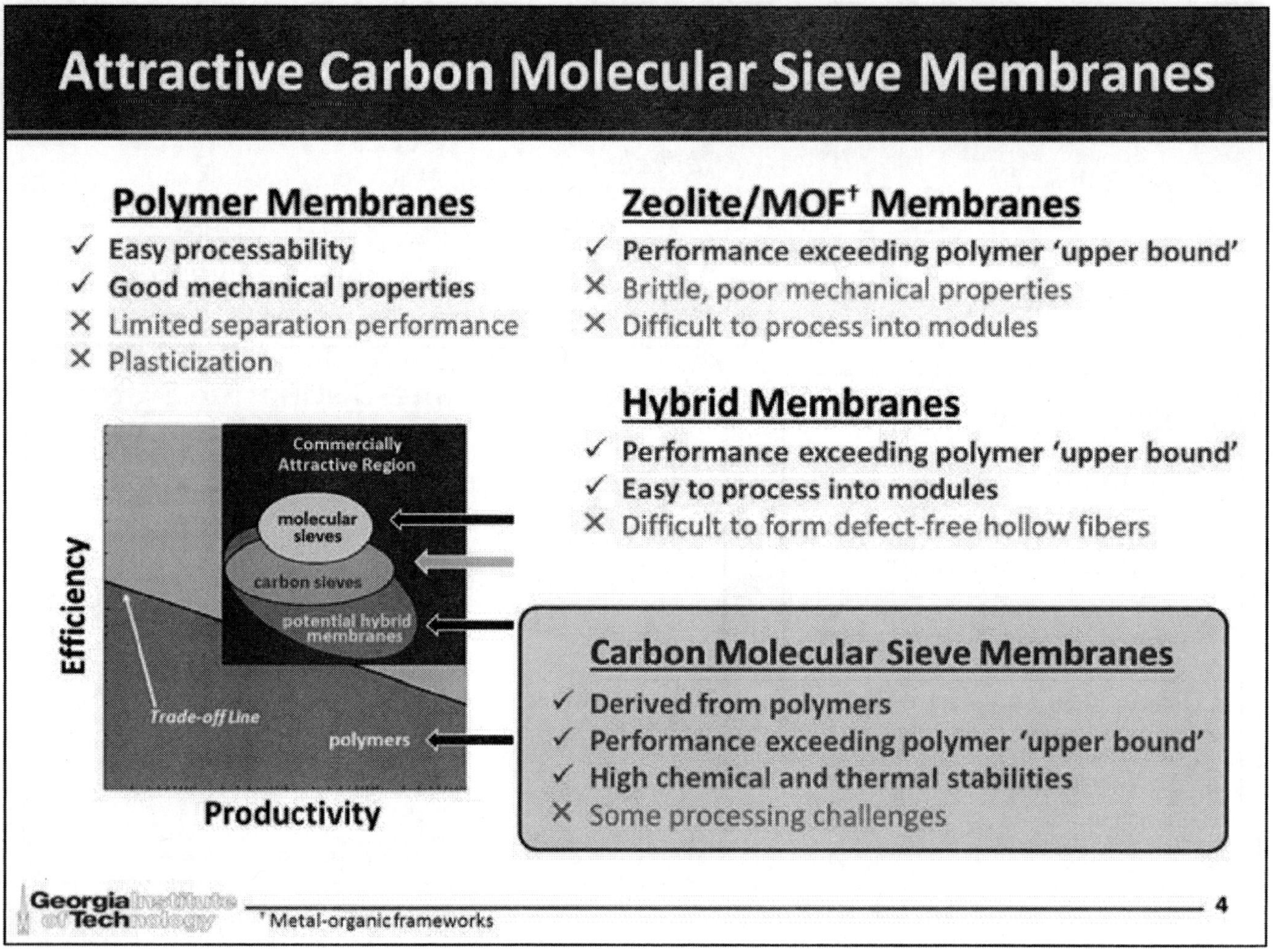

- There are several categories for gas membrane materials.
- This work focuses on the study of attractive material: carbon molecular sieve (CMS) membrane.

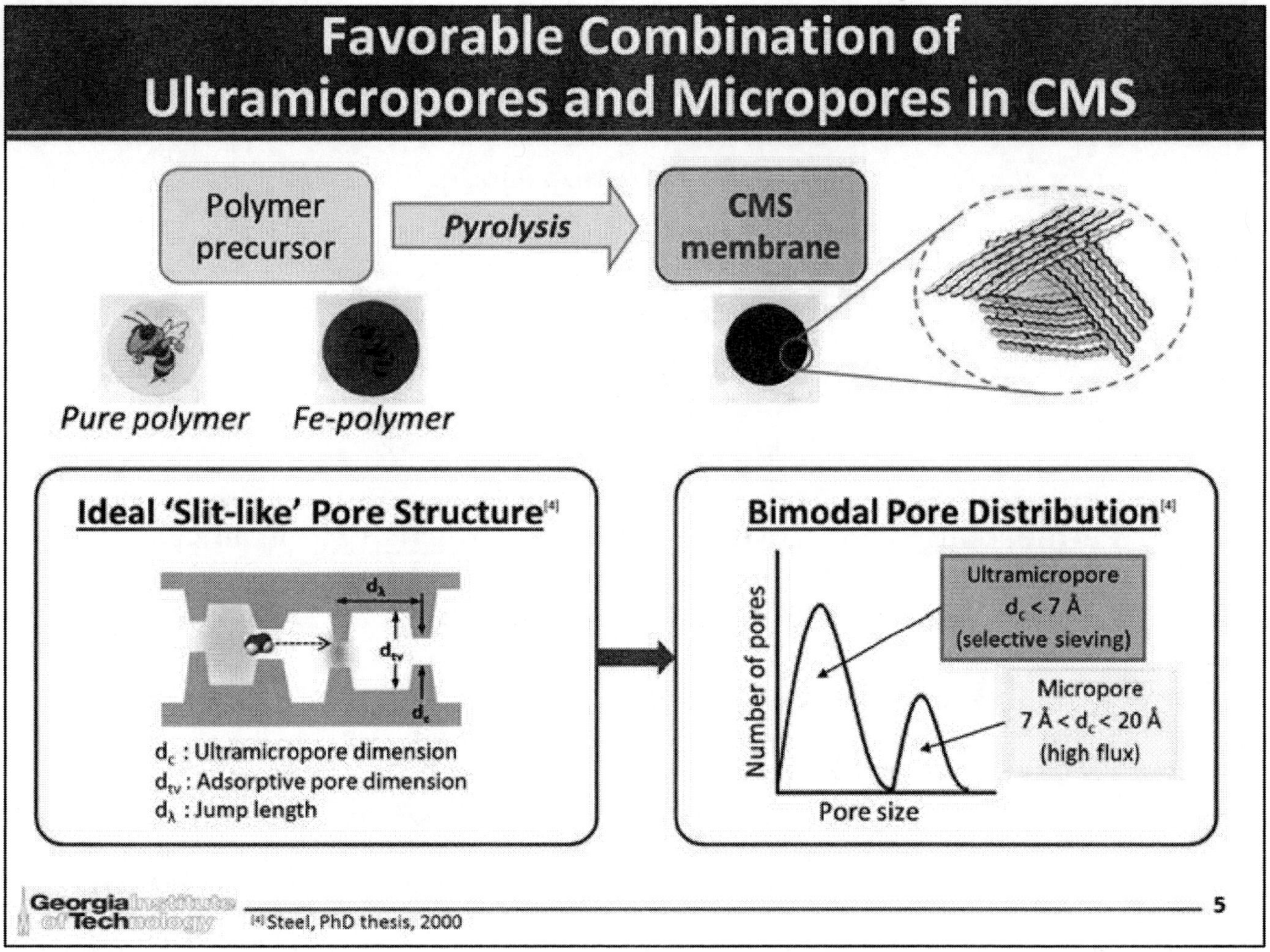

- A CMS membrane can be formed by thermal decomposition of a polymer precursor at high temperature.
- The combination of rigid ultramicropores (providing selective molecular sieving) and micropores (providing high productivity and sorption sites) makes CMS membranes an attractive material for gas separations.

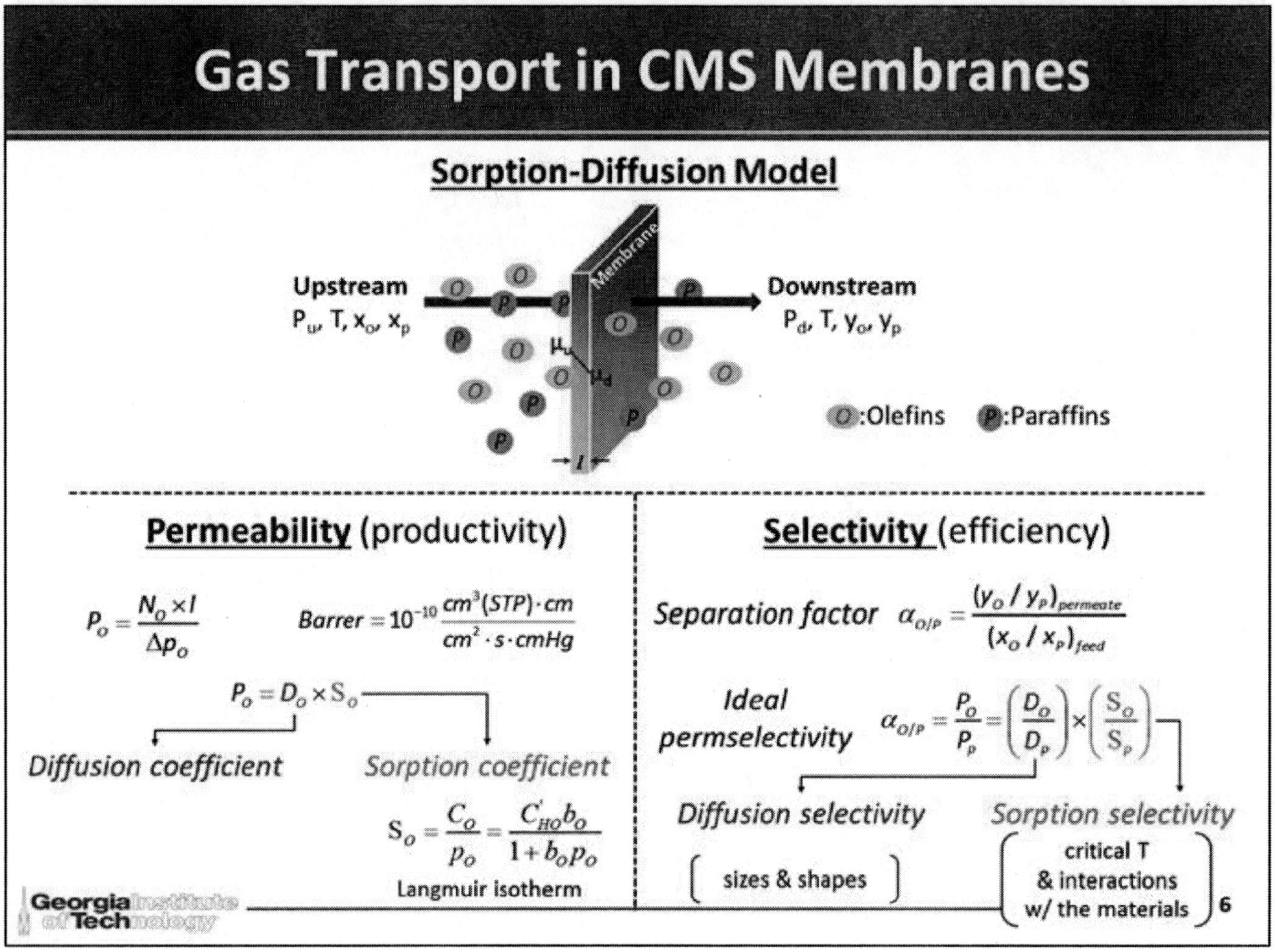

- Gas transport through a CMS membrane can be described by sorption-diffusion model.
- Permeability and selectivity are two intrinsic properties used to characterize membrane performance.

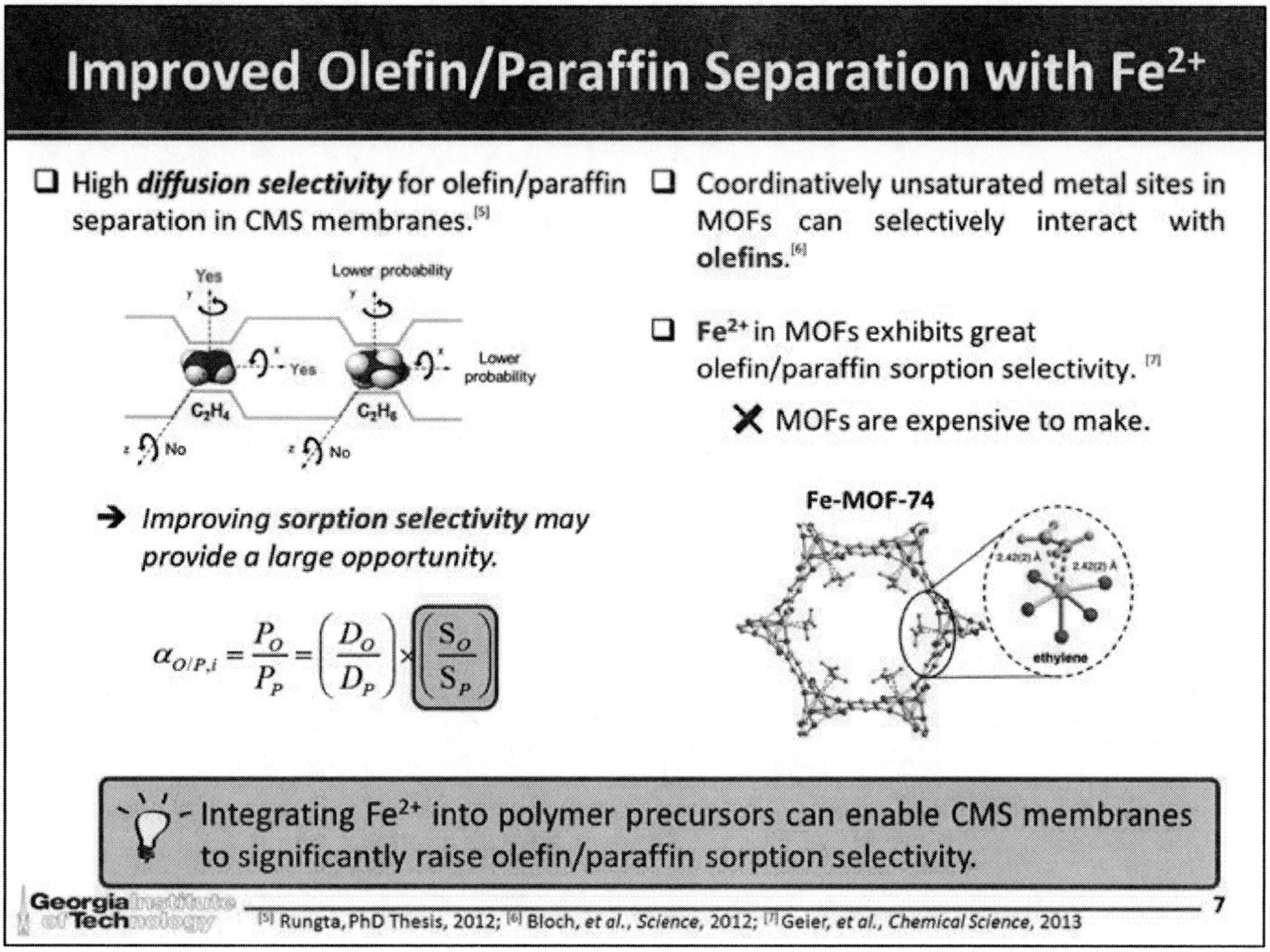

- Integrating Fe^{2+} into polymer precursors was pursued to improve CMS membrane separation properties for advanced olefin/paraffin separations.
- The choice of Fe^{2+} was based on a combination of factors including olefin selectivity, low cost, and stability against contaminants.

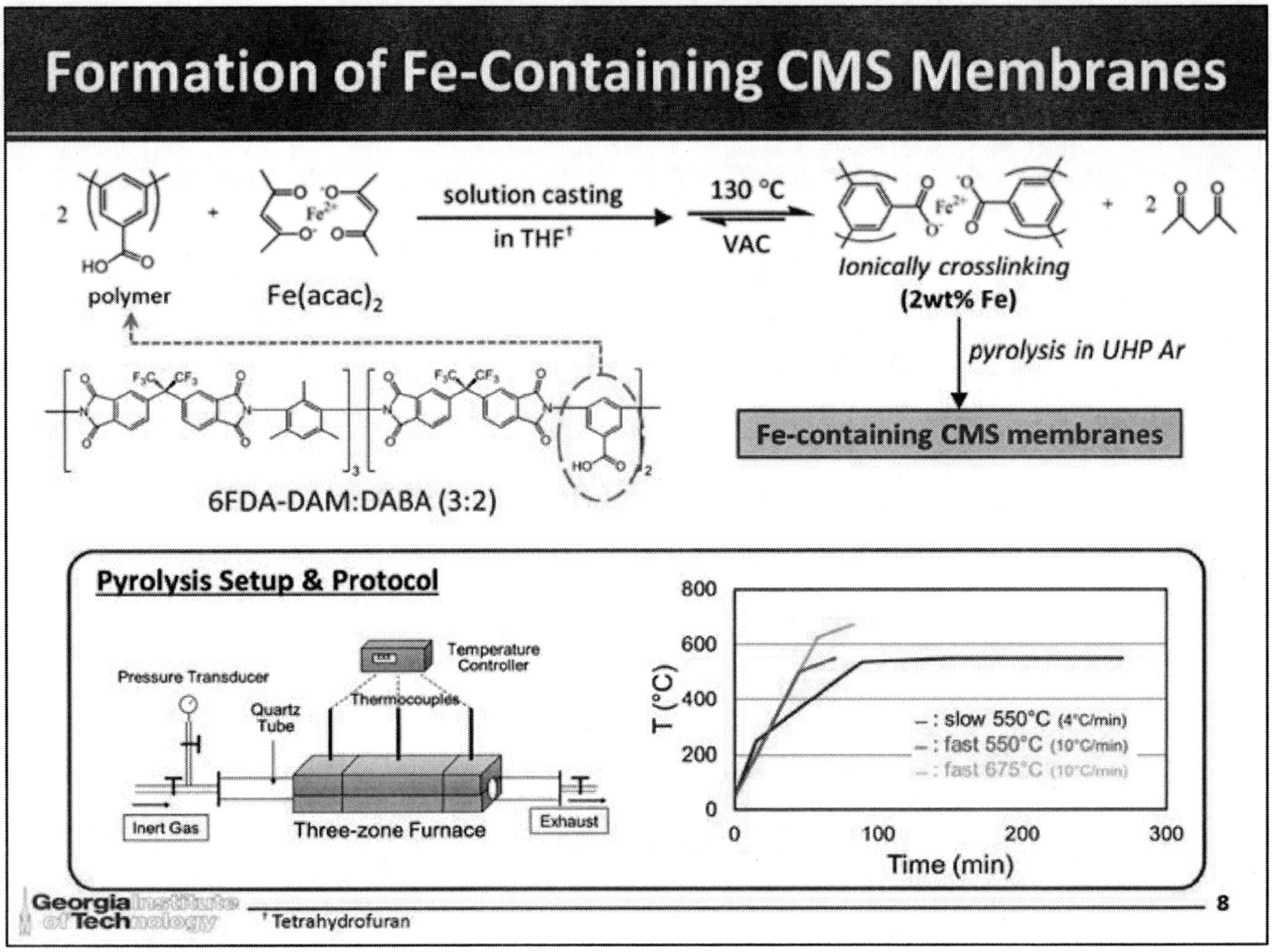

- The Fe^{2+} can be cross-linked with the carboxyl group in the polyimide, 6FDA-DAM:DABA (3:2), to have better distribution.
- Three different pyrolysis protocols were proposed to study the effect of oxidation state of Fe on separation performance.

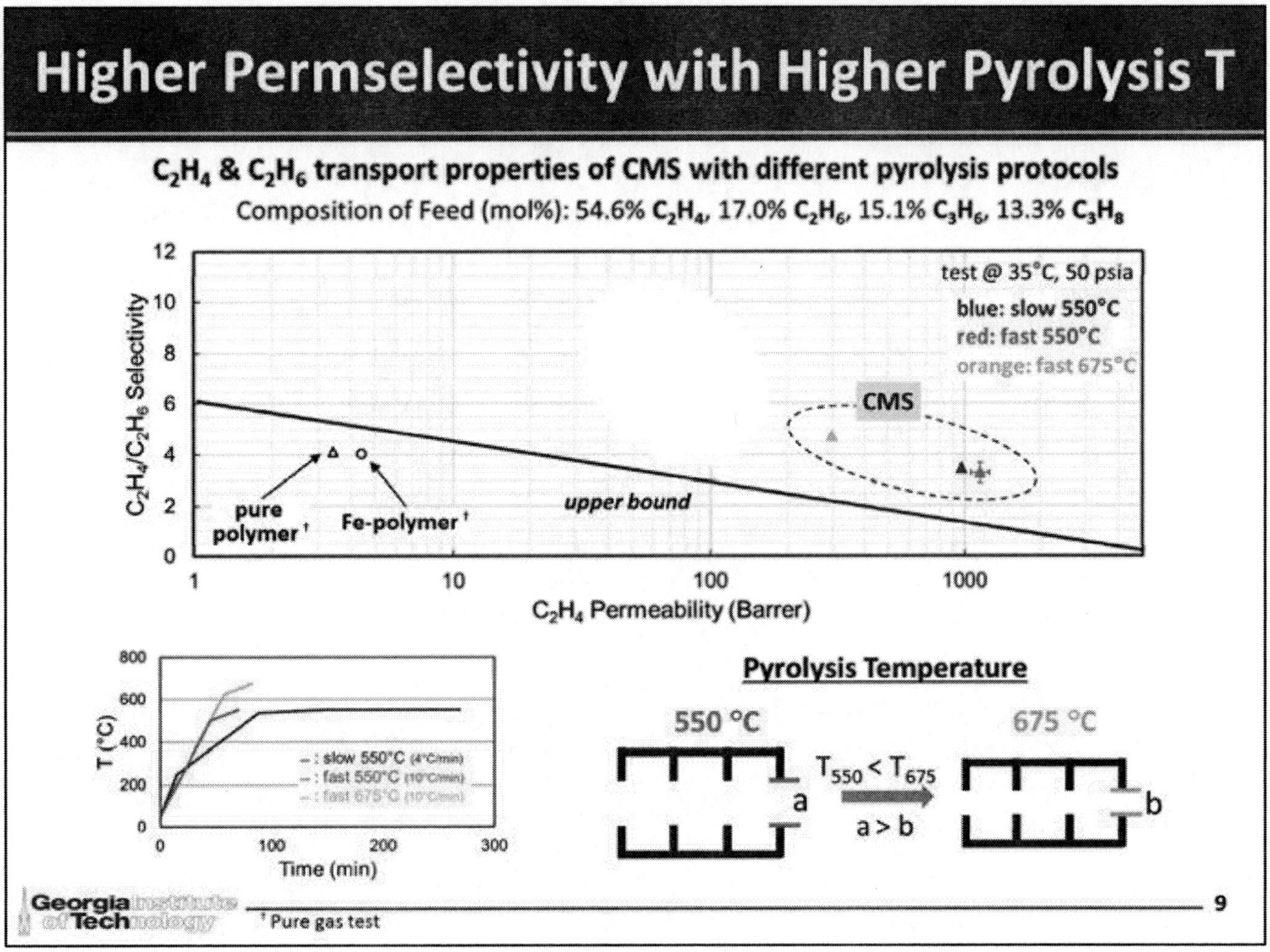

- CMS membranes derived from pure polymer have performance above upper bound.
- Both micropores and ultramicropores are tightened at high pyrolysis temperature, and that results in an increase in selectivity and a decrease in permeability.

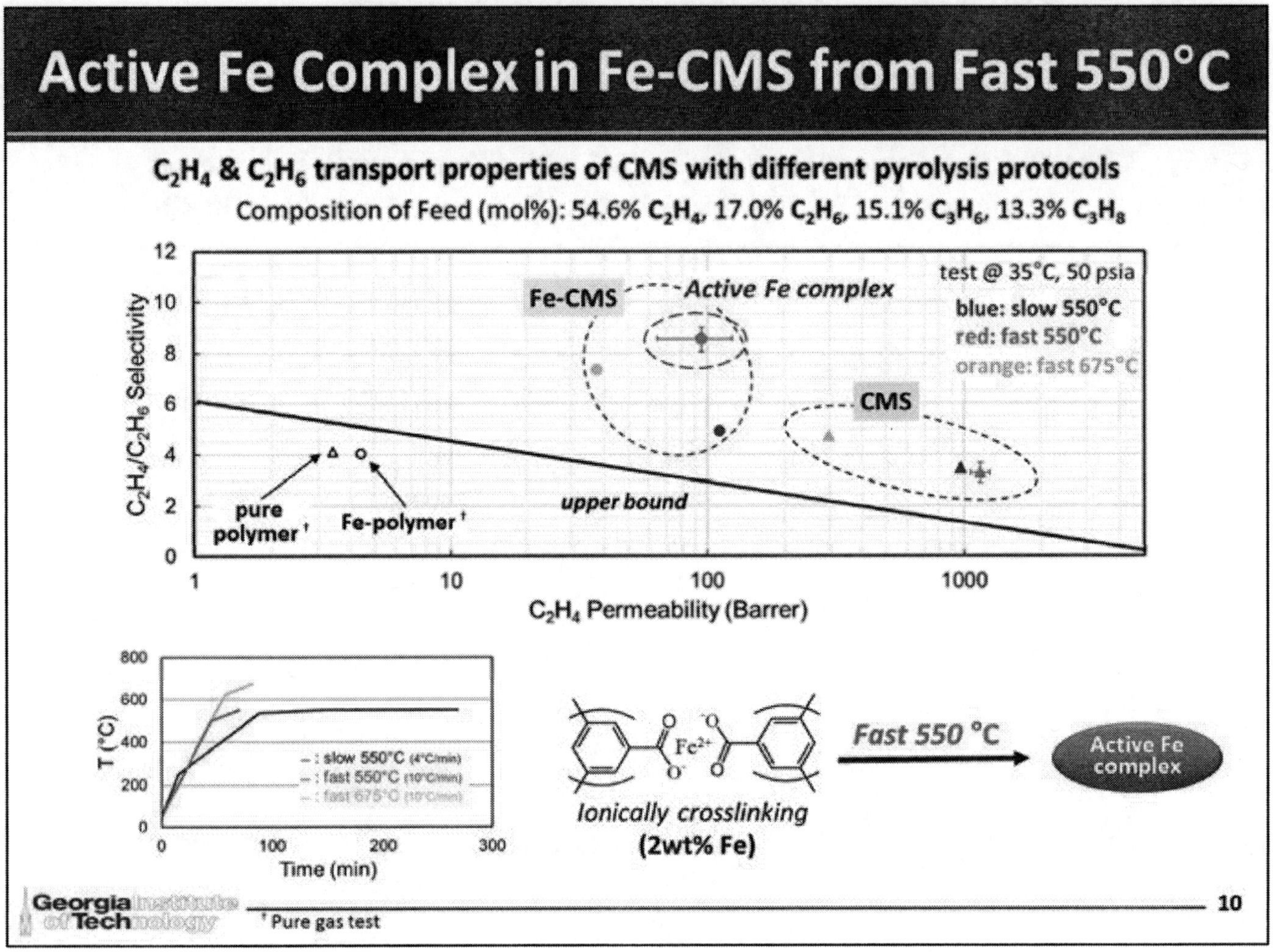

- As for Fe-CMS, Fast 550°C leads to highest selectivity
- Lower pyrolysis temperature and faster heating rate may generate specific active Fe complex which have selective interaction with olefins.
- Although Fe causes a reduction in permeability, Fe-CMS still have promising performance compared to other separation materials.

Desirable Valence State of Fe Higher than 0

Fe K-edge XANES† spectra of Fe-CMS membranes

(data collected by David F. Yancey at Argonne Natl Lab)

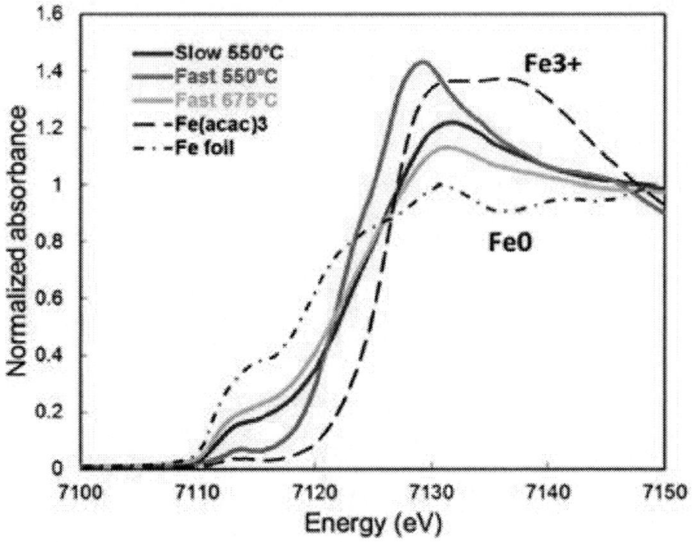

Fe-CMS membranes have improved performance ($\alpha_{C_2H_4/C_2H_6} > 9$) when applying Fast 550°C protocol.

- Fast 550°C results in an average Fe bonding state that is closer to Fe^{3+}.
- The valence state of Fe higher than 0 is desirable for recovery of olefins.

Georgia Institute of Technology

11

†X-ray absorption near edge structure

- X-ray absorption near edge structure (XANES) spectra show that valence state of Fe higher than 0 is desirable for recovery of olefins.

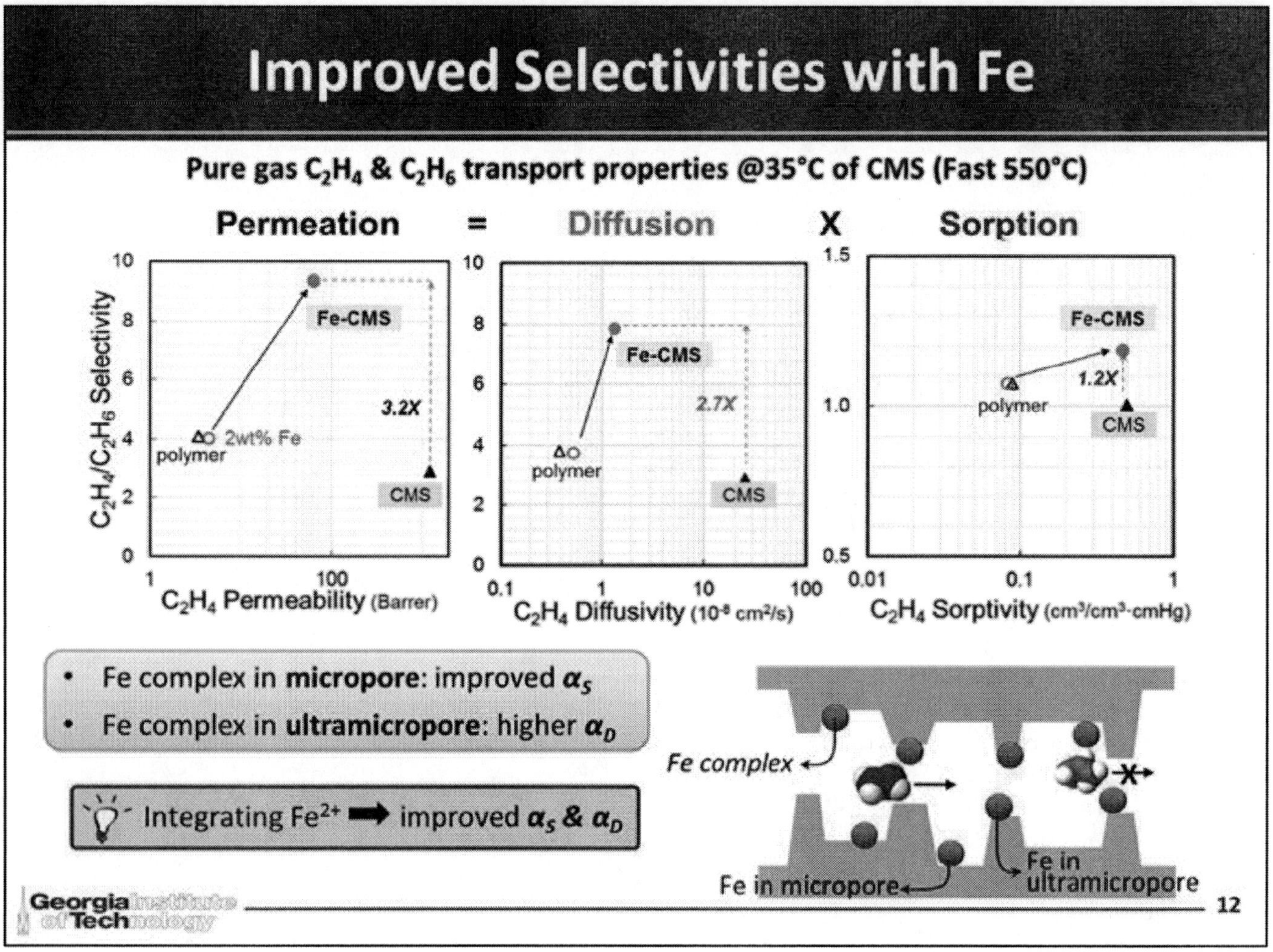

- Fe complex can not only present in micropores to increase sorption selectivity but also in ultramicropores to increase diffusion selectivity.

Hypothetical Effects of Fe in CMS

- Fe complex in **micropore**:
  - Reduced paraffins sorption capacity
  - Improved α_S based on higher affinity between Fe & olefins
- Fe complex in **ultramicropore**:
  - larger CMS ultramicropores are blocked
  - higher α_D

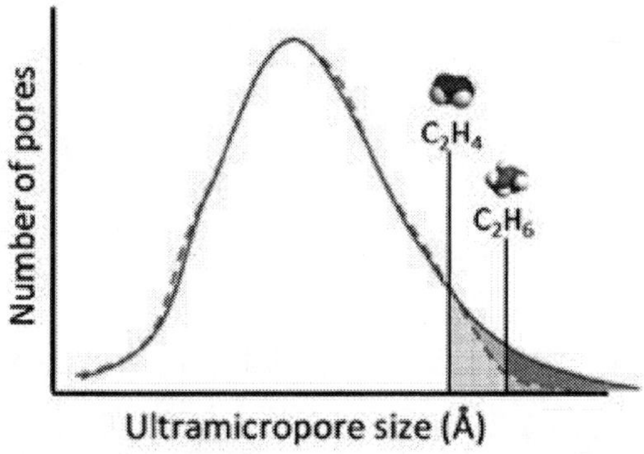

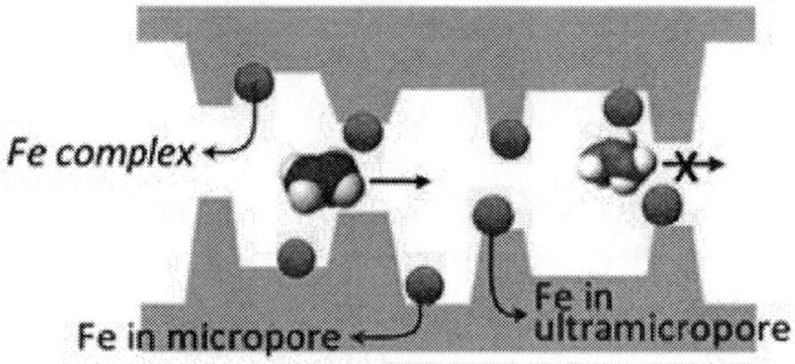

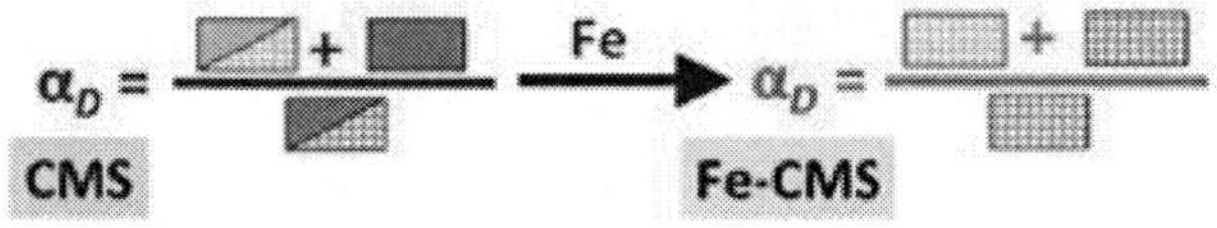

Georgia Institute of Technology 13

- Fe complex in micropores can have higher affinity with olefins and leads to improved sorption selectivity.
- Fe complex can blocked larger ultramicropores, and it has more effects on bulky paraffins. Hence, diffusion selectivity increases.

Stable Fe-CMS with Active Feed

Gas transport properties @35°C of Fe-CMS (Fast 550°C), quaternary feed

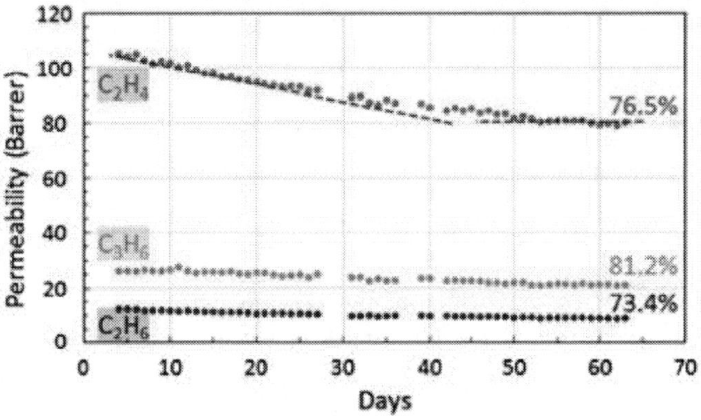

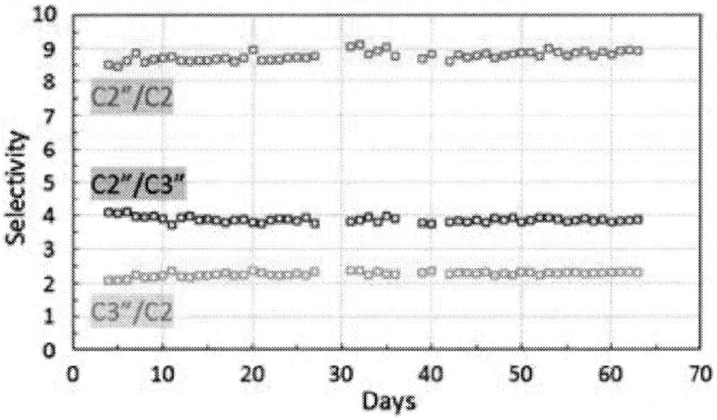

Composition of feed & permeate at 63rd day

Gas	Feed (mol%)	Permeate (mol%)
C_2H_4	54.6	90.33
C_3H_6	15.1	6.42
C_2H_6	17.0	3.15
C_3H_8	13.3	0.05

Olefins > 96.7%

- Fe-CMS is stable after 50 days with active feed
- Reduced ***P***: relaxation pores in CMS
- Stable ***α***: stably active Fe complex

Georgia Institute of Technology 14

- Fe-CMS membrane is stable after 50 days with a active feed.
- Olefins can be enriched higher than 96% in permeate side.

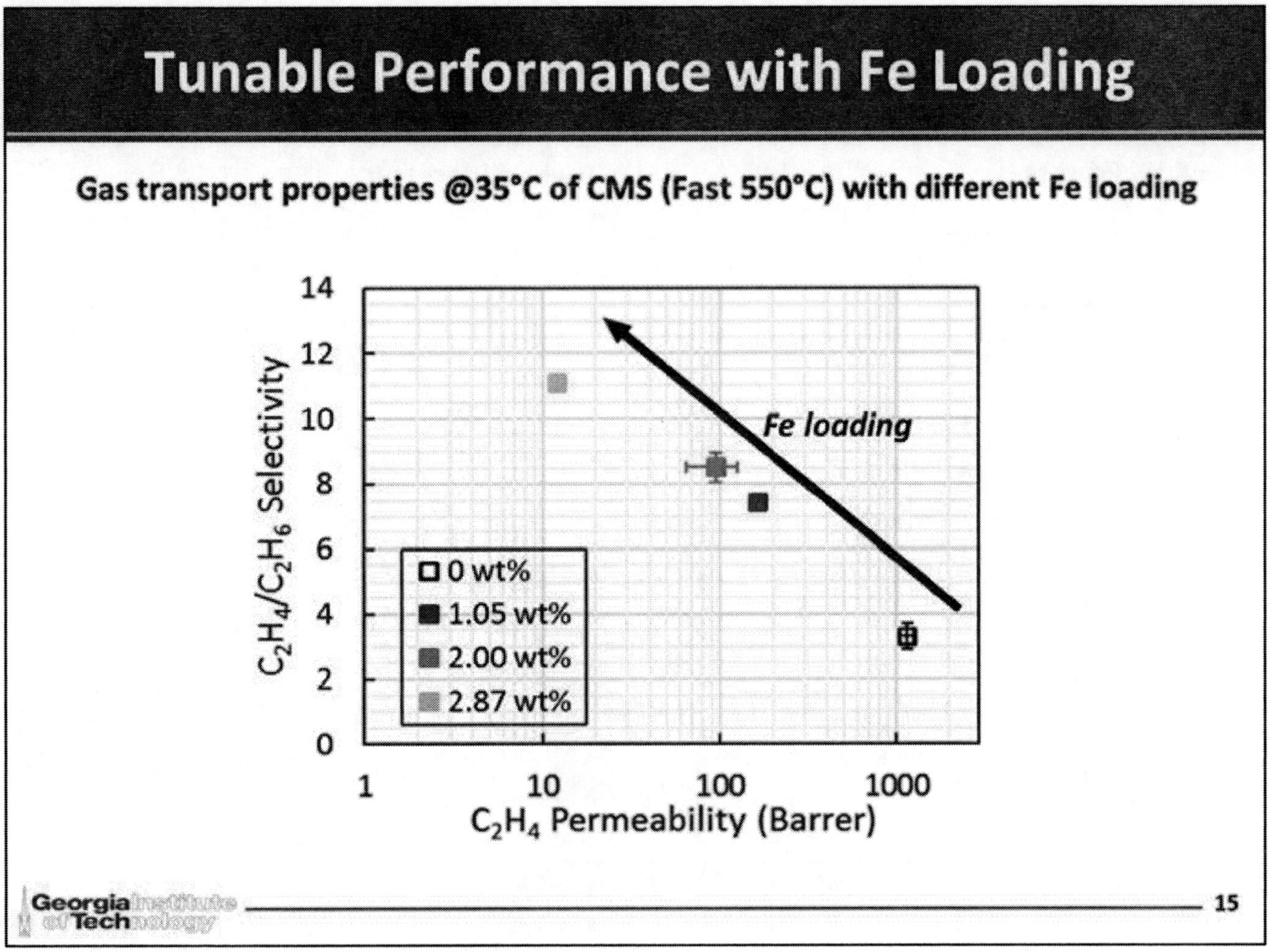

- The separation performance of Fe-CMS membranes can be tuned by the loading of Fe in the precursors.

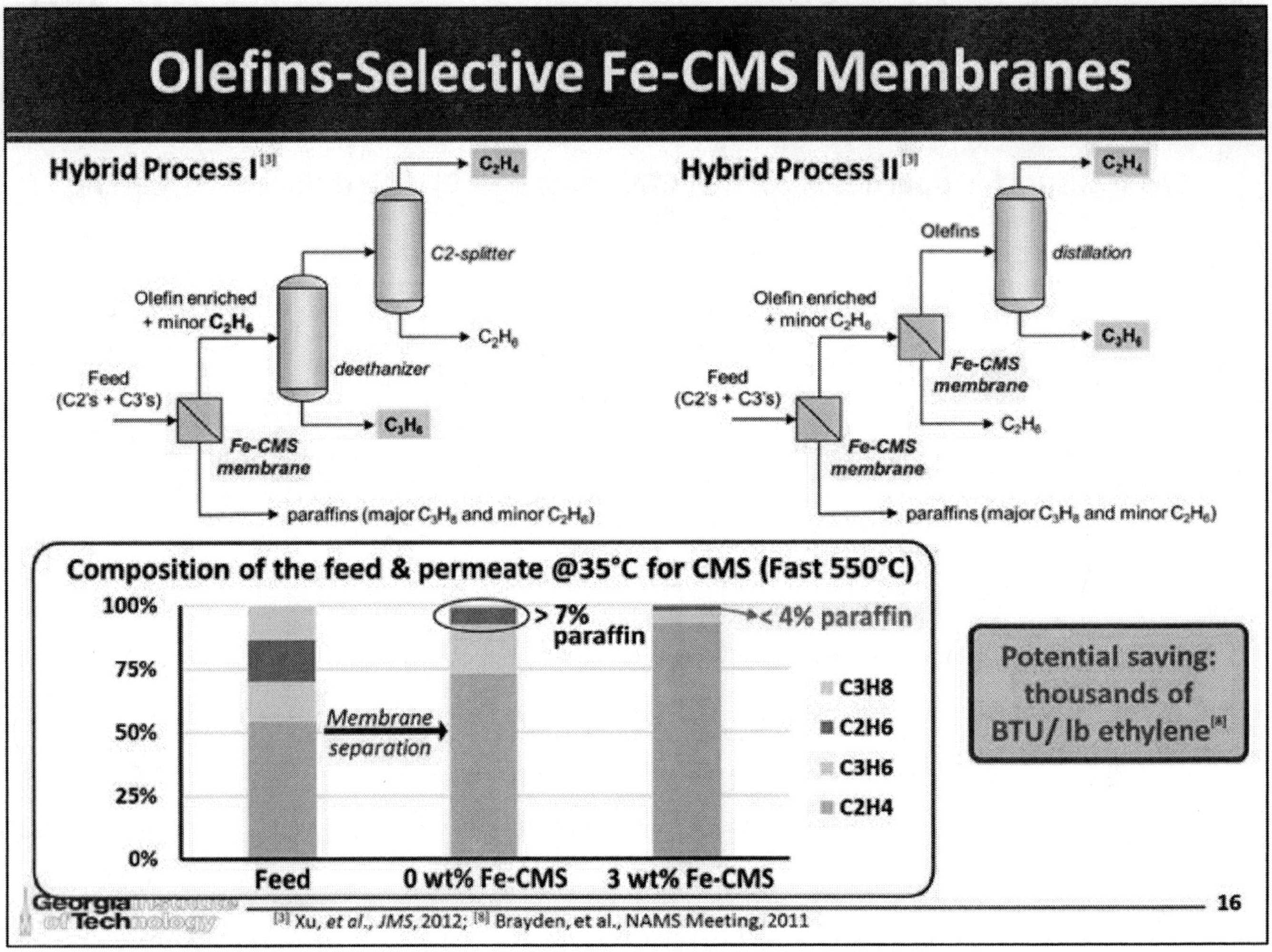

- The Fe-CMS membrane is olefins-selective and enriches permeate side with 97% olefins.
- The application of the olefins-selective membranes can retrofit the current distillation separation processes and achieve significant energy savings.

Conclusions

- Incorporation of Fe^{2+} in CMS membranes can significantly improve separation performance of olefins/paraffins when applying lower pyrolysis temperature and faster heating rate (i.e., Fast 550°C).
- XANES was performed to characterize Fe in Fe-CMS, and it shows that Fe with valence state higher than 0 is desirable for recovery of olefins.
- Fe-CMS is stable with active feed, and the performance is attractive.
- Performance of Fe-CMS can be tuned with Fe loading to achieve olefins-selective membranes.

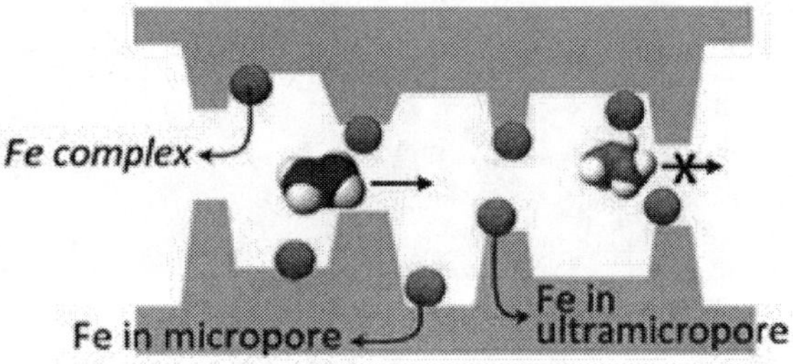

Georgia Institute of Technology 17

Acknowledgements ☺

- Koros Research Group at Georgia Tech
- The Dow Chemical Company

Georgia Institute of Technology

18

Thank You

QUESTIONS & COMMENTS

Yu-Han Chu
Email: yuhanchu@gatech.edu

ID# 479866

Combining Add-On Technologies for GSP Plants with up-to-date Plate Fin Heat Exchanger Design

Why It Makes Sense To Retrofit A Plant When The Feed Composition Changes

Thomas Walter
Thomas Martin
Dr. Adrian Wegmann
Linde Engineering North America, Inc.

THE LINDE GROUP
Linde

History Of The GSP Plant

THE LINDE GROUP
Linde

- The Gas Subcooled Process was developed in the late 1970s based on the simple turbo expander process
- For many years this process was the cutting edge technology in natural gas processing
- GSP plants can provide ethane recoveries up to 95% in ethane recovery mode
- GSP plants can provide propane recoveries up to 98% in propane recovery mode

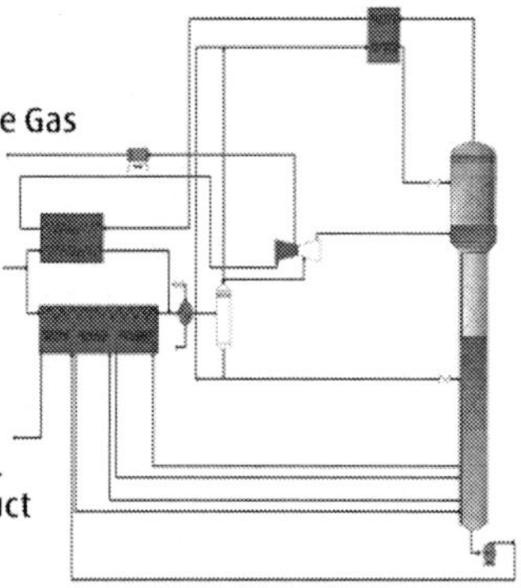

Also see GAS Processors Suppliers Association Engineering Data Book 13th Edition (2012) "GSP Process for Propane Recovery" pp. 16-14

How Can A Small Composition Change Affect Our Plant?

THE LINDE GROUP

Linde

- Any change in composition, temperature, or pressure has the potential to significantly impact a gas plant's performance

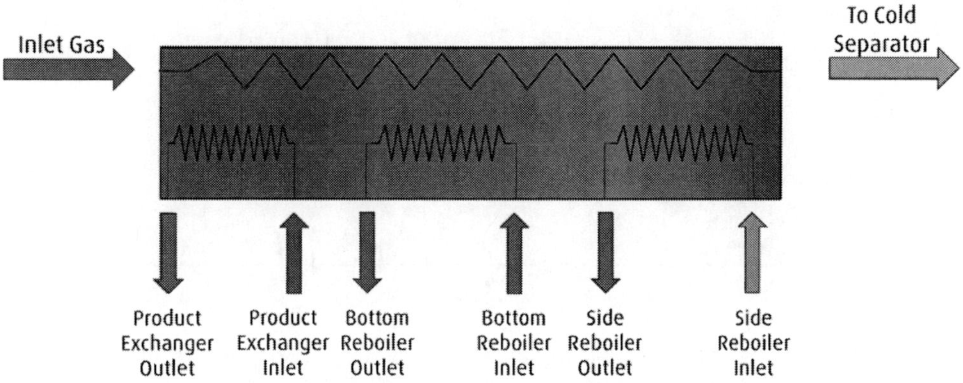

How Can A Small Composition Change Affect Our Plant?

THE LINDE GROUP

Linde

- Any change in composition, temperature, or pressure has the potential to significantly impact a gas plant's performance

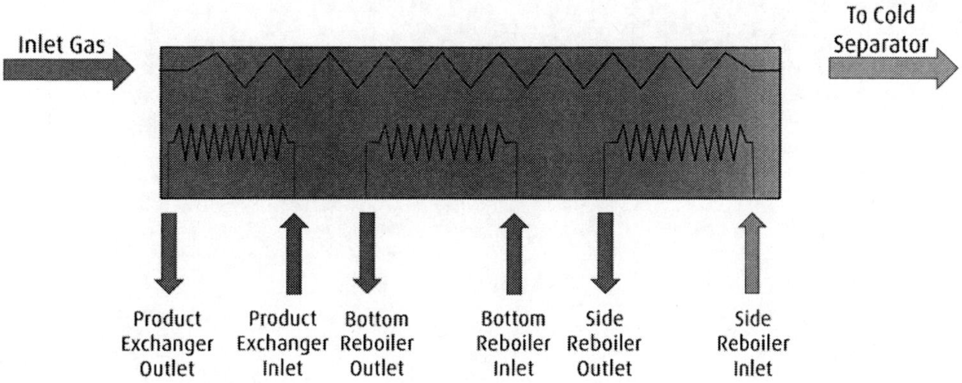

Market Outlook For Ethane Recovery

- The downturn in the price of ethane has led companies to purchase plants primarily optimized for ethane rejection
  - The price of ethane is expected to greatly increase in late 2017 when the ethane crackers on the Gulf Coast come online
  - Plants optimized for ethane rejection only will not have capability to maximize profits when operating in ethane recovery
  - The GSP design does not have a wide range of flexibility to handle changing feed conditions

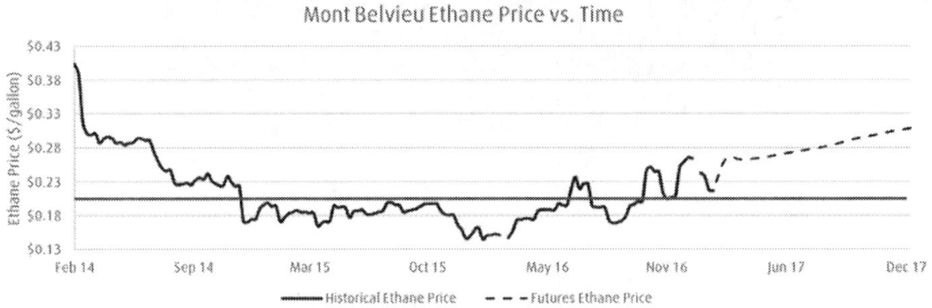

Historical Pricing: https://www.quandl.com/data/CHRIS/CME_C01-Mont-Belvieu-Ethane-OPIS-Futures-Continuous-Contract-1-C01-Front-Month?utm_medium=graph&utm_source=quandl Futures Pricing: Obtained from CME Group Mont Belvieu Ethane Pricing

What Is Preventing High Ethane Recoveries In Our Plant?

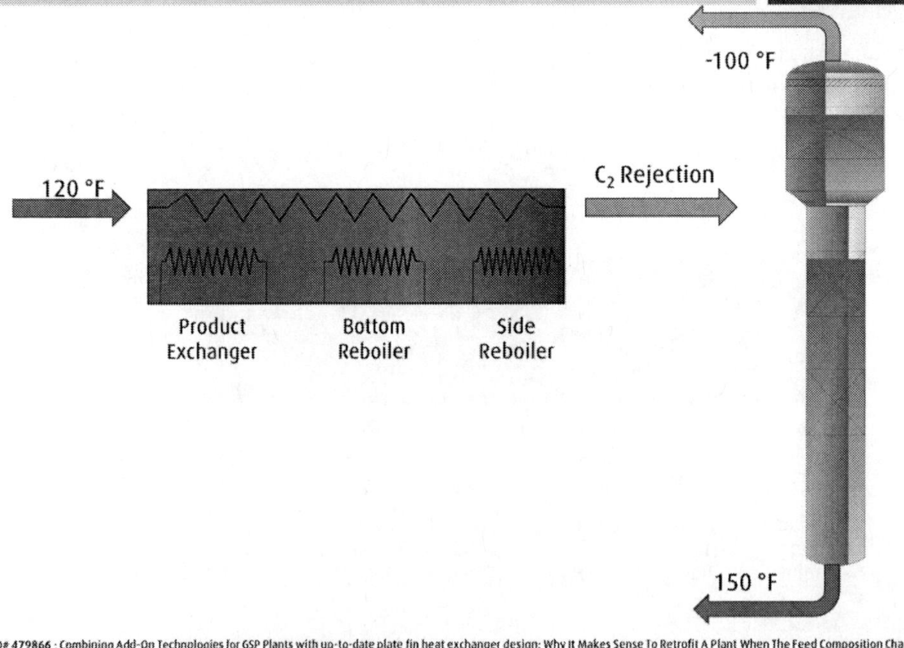

What Is Preventing High Ethane Recoveries In Our Plant?

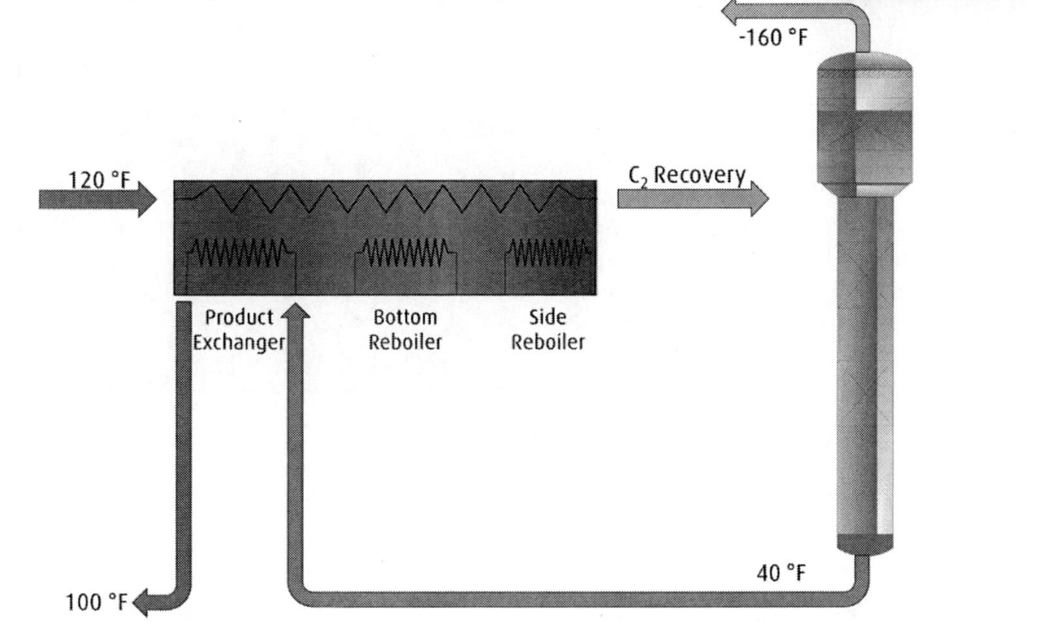

What Is Preventing High Ethane Recoveries In Our Plant?

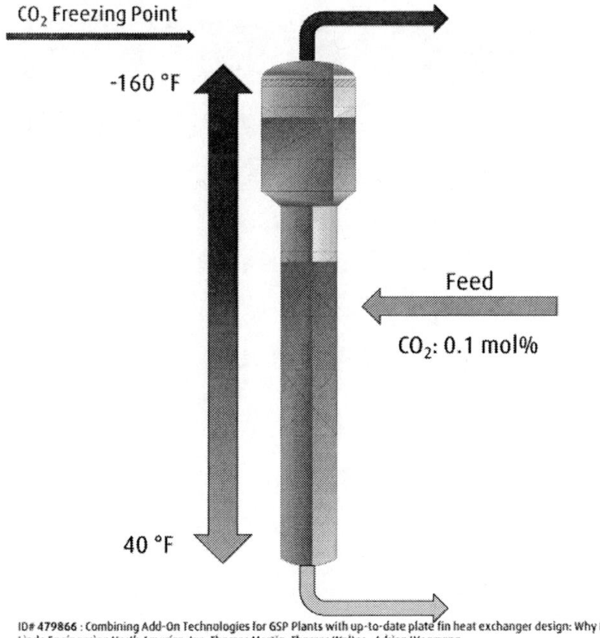

- For plants experiencing increasing CO_2 concentrations in their GSP plant, solid formation can occur at high ethane recoveries
  - A plant may have the ability to recover 95%+ C_2, but if the concentration of CO_2 is high enough, CO_2 freezing in the top of the column limits the plant's performance
  - Adding a recycle of lean gas with low concentrations of CO_2 to the top of the column can lower the concentration of CO_2, thus lowering the freezing temperature to allow for higher ethane recoveries

How Can We Increase Throughput In Our Plant?

- Gas plants are limited by the flow to the expander, refrigeration capacity, or the demethanizer liquid/vapor traffic
- These are all very costly units which don't make economic sense to replace
- Through proper heat integration and rigorous rating software, a plant's capacity can be increased by adjusting process parameters such as column pressure and reboiler duties
- The plant capacity can also be increased by replacing either of the plate-fin heat exchangers with an optimized exchanger

Design Conditions

Specifications and Feed conditions:

- Flow rate 200 MMscfd
- Temperature 92.8 F
- Feed Pressure 780 psig
- Residue Pressure 1100 psig
- NGL product Pressure 1440 psig
- C1/C2+ Liq. vol 0.50%
- C1/C2 Liq. vol 1.50%
- CO2/C2 Liq. vol 0.35%

Component	Design Composition	New Composition
Carbon Dioxide	0.0060	0.0059
Nitrogen	0.0040	0.0036
Methane	0.8461	0.8174
Ethane	0.0930	0.1098
Propane	0.0273	0.0350
Butanes and Heavier	0.0236	0.0283
GPM	**4.06**	**4.89**

GSP - Base Design

Pros:

- Lowest initial capital cost
- Standardized design available
- Operates well in C_2 Rejection mode (>97% C_3 Recovery)

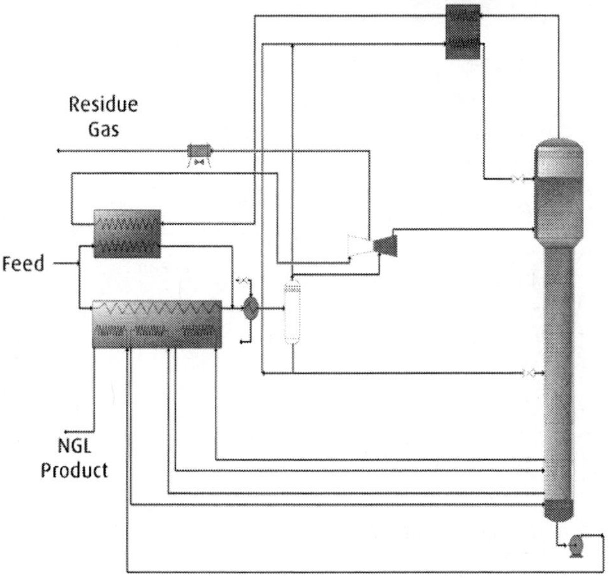

Cons:

- Ethane recovery limited to 88.5%
- Plant is bottlenecked
- Technology is prone to CO_2 freezing at high ethane recoveries

Retrofit Option 1 - GSP With Shell And Tube Exchanger

Pros:

- Low cost to install exchanger
- Minor modifications to piping
- Ethane recovery is improved to 93.2% or
- Plant capacity can be increased to 216.9 MMscfd
- Modification possible within short shut-down time

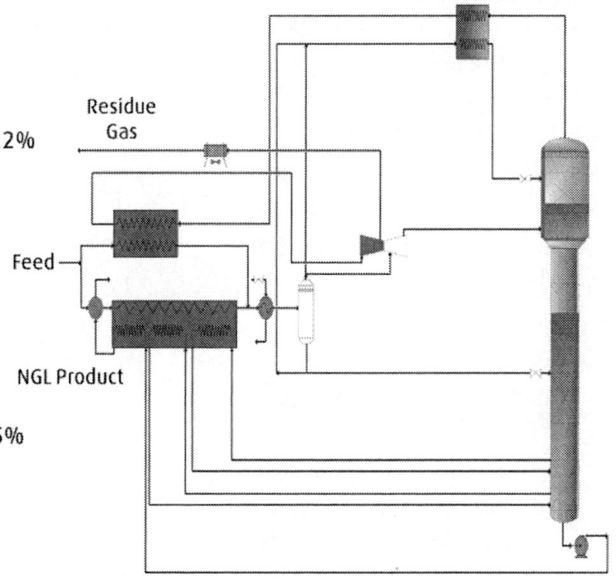

Cons:

- Very high ethane recoveries (>95% C_2 Recovery) and plant capacity above 220MMscfd not possible

Retrofit Option 2 - GSP With New PFHE

THE LINDE GROUP

Linde

Pros:

- Ethane recovery is improved to 94.3% or
- Plant capacity can be increased to 224.5 MMscfd
- Identical to original process set-up
- Low costs and modification possible within short shut-down time

Cons:

- Limited by CO_2 freezing at high ethane recoveries (>95% C_2 Recovery)
- Medium capital costs

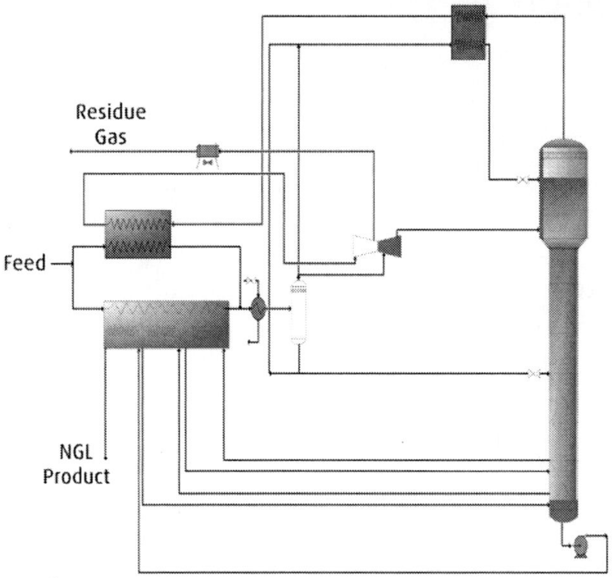

ID# 479866 : Combining Add-On Technologies for GSP Plants with up-to-date plate fin heat exchanger design: Why It Makes Sense To Retrofit A Plant When The Feed Composition Changes 13
Linde Engineering North America, Inc. Thomas Martin, Thomas Walter, Adrian Wegmann

Retrofit Option 3 - Residue Recycle

THE LINDE GROUP

Linde

Pros:

- Ethane Recovery is improved to 95.2% or
- Plant capacity can be increased to 216.9 MMscfd
- CO_2 tolerant design
- Provides greater flexibility for changing feed conditions

Cons:

- Medium to high capital costs and longer installation period
- New process for operators to learn
- Less potential to increase capacity compared to improved recovery

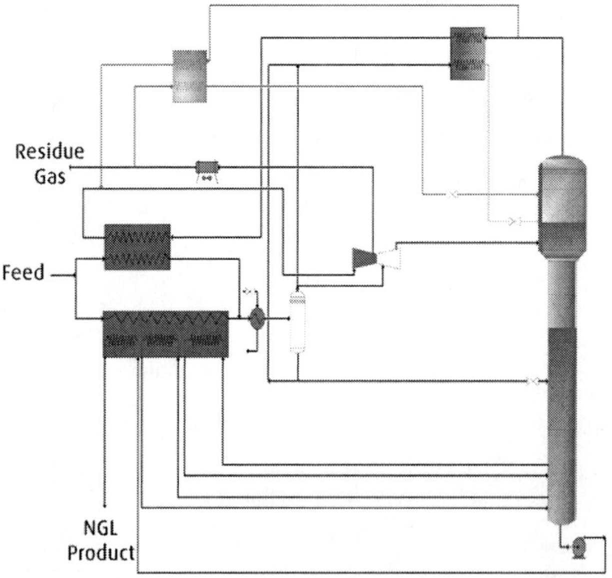

ID# 479866 : Combining Add-On Technologies for GSP Plants with up-to-date plate fin heat exchanger design: Why It Makes Sense To Retrofit A Plant When The Feed Composition Changes 14
Linde Engineering North America, Inc. Thomas Martin, Thomas Walter, Adrian Wegmann

Retrofit Option 4 - Residue Recycle With New PFHE

THE LINDE GROUP

Linde

Pros:

- Ethane Recovery is improved to 97.7% or
- Plant capacity can be increased to 220.9 MMscfd
- CO_2 tolerant design
- Fully optimized design

Cons:

- Highest capital costs and longer installation period
- New process for operators to learn
- Less potential to increase capacity compared to improved recovery

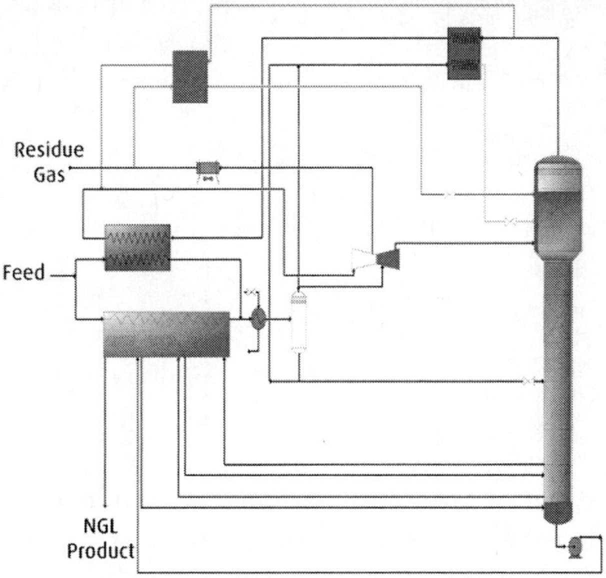

Case Study

THE LINDE GROUP

Linde

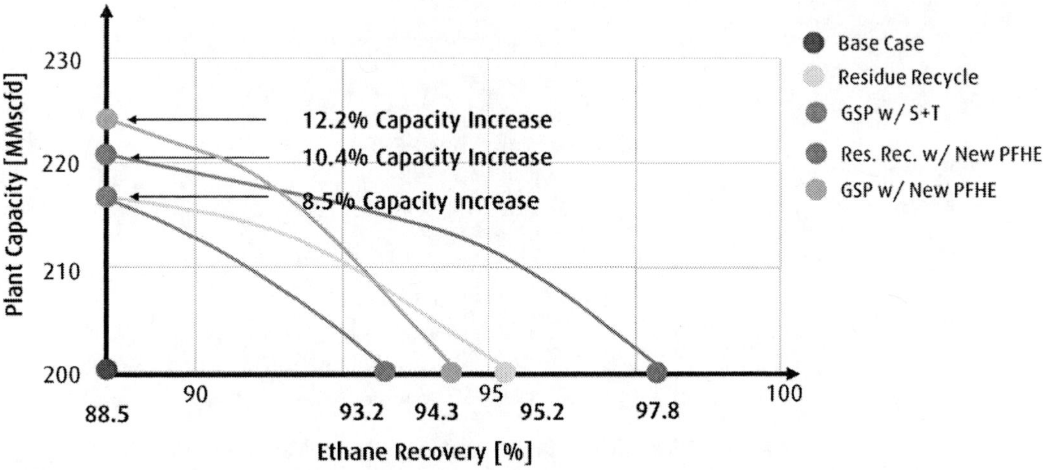

→ **Ethane recovery and/or plant capacity can improved depending on client needs**

Economics For Increased Ethane Recovery

	GSP with Shell and Tube	GSP with New PFHE	Residue Recycle	Residue Recycle with New PFHE
Modifications:	- S&Tube Hex (NGL Product)	- Warm Gas/Gas PFHE	- Subcooler PFHE - DeMethanizer - Residue compression	- Subcooler & Gas/Gas PFHE - DeMethanizer - Residue compression
Total costs:	Low (Base)	Low-Moderate (~3 times Base)	Moderate – High (~6 times Base)	High (~8 times Base)
C_2 price Break even 1-year:	$0.26-0.28	$0.32-0.42	$0.46-0.56	$0.54-0.64
C_2 price Break even 2-years:	$0.23-0.24	$0.34-0.39	$0.33-0.38	$0.37-0.43

→Retrofits may be economical even at lower Ethane prices

→Economic gains are more significant if increase in plant capacity is considered

1) Ethane valued at $0.20/gal as heating value. Fractionation and transportation costs not included

Concept overview

GSP with Shell and Tube

+
- Short installation time
- Lowest cost/shortest payback period

−
- Very high ethane recovery or capacity increase not possible
- Prone to CO_2 freezing

GSP with New Warm Gas PFHE

+
- Provides highest capacity increase and high C_2 recovery
- Short installation time

−
- Prone to CO_2 freezing
- Medium investment/longer payback time

Recycled Residue Gas Layout

+
- Not prone to CO_2 freezing
- High CO_2 recovery

−
- Higher investment cost
- Long installation time
- New process configuration for operators

Recycled Residue Gas Layout with New Warm Gas PFHE

+
- Highest recovery increase
- Not prone to CO_2 freezing

−
- Highest investment cost
- Long installation time
- New process configuration for operators

Summary

- Plants with changed inlet conditions and standard plants can be improved significantly by retrofits
- The various retrofit options require deep knowledge of the process and equipment design (e.g. PFHE's) to identify the best solution for the client
- Even at low to moderate C_2 prices, retrofits may be economical with a short payback time
- Retrofits can be designed to increase recoveries, plant capacity or a combination of both to maximize client profit

→ Optimal solution can be developed in cooperation between client and technology provider

Thanks for your attention.

Vielen dank für Ihre Aufmerksamkeit.

Thomas.Walter@Linde.com

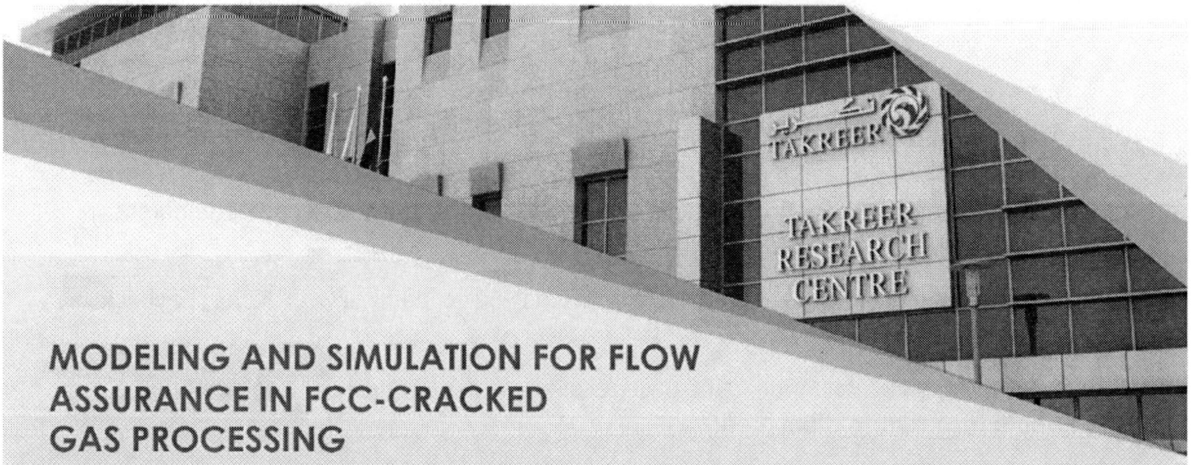

MODELING AND SIMULATION FOR FLOW ASSURANCE IN FCC-CRACKED GAS PROCESSING

M. KARAKAYA, M. RAKIB,
M. MUSHARFY, M. SULEIMAN

AIChE 2017 Spring Meeting and
13th Global Congress on Process Safety

شركة أبوظبي لتكرير النفط (تكرير)
مركز تكرير للأبحاث
Abu Dhabi Oil Refining Company (TAKREER)
TAKREER Research Centre (TRC)

TAKREER
We Refine Right

PRESENTATION CONTENT

- Process Background
- Focus on the Deethanizer Circuit
- Process & Hydrate Formation Model
- Prevention of Hydrate Formation
- Key Findings and Conclusions

AIChE 2017 Spring Meeting and
13th Global Congress on Process Safety
2

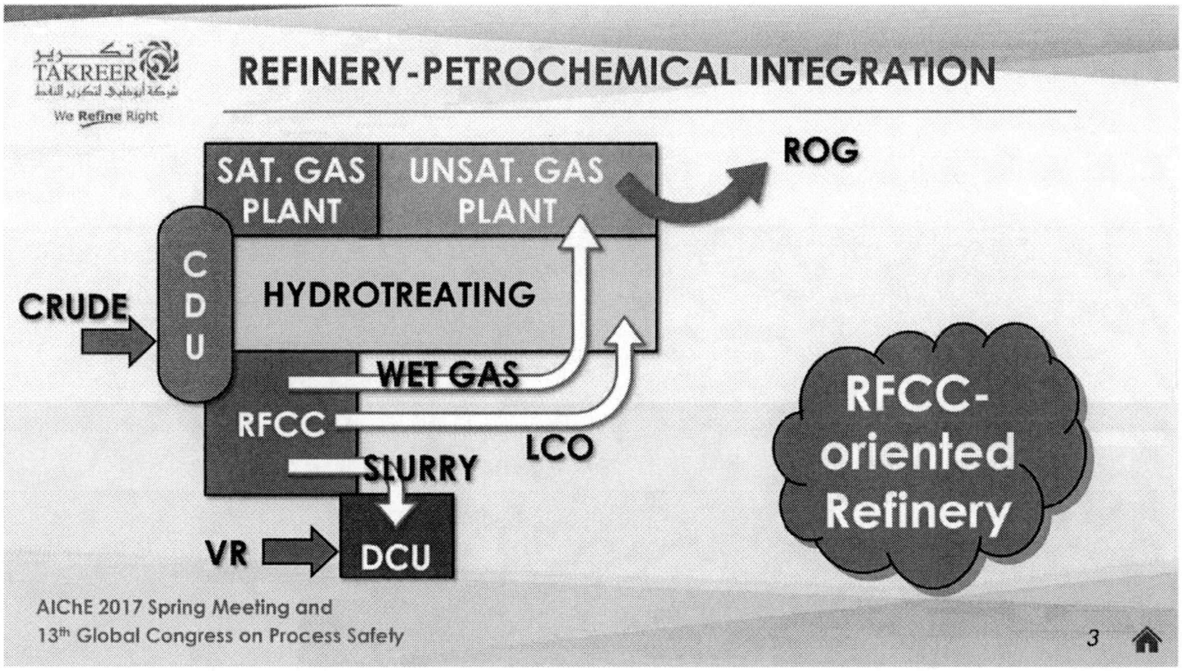
TAKREER
We Refine Right
REFINERY-PETROCHEMICAL INTEGRATION
SAT. GAS PLANT
UNSAT. GAS PLANT
ROG
CDU
CRUDE
HYDROTREATING
WET GAS
RFCC
LCO
SLURRY
VR
DCU
RFCC-oriented Refinery
AIChE 2017 Spring Meeting and
13th Global Congress on Process Safety
3

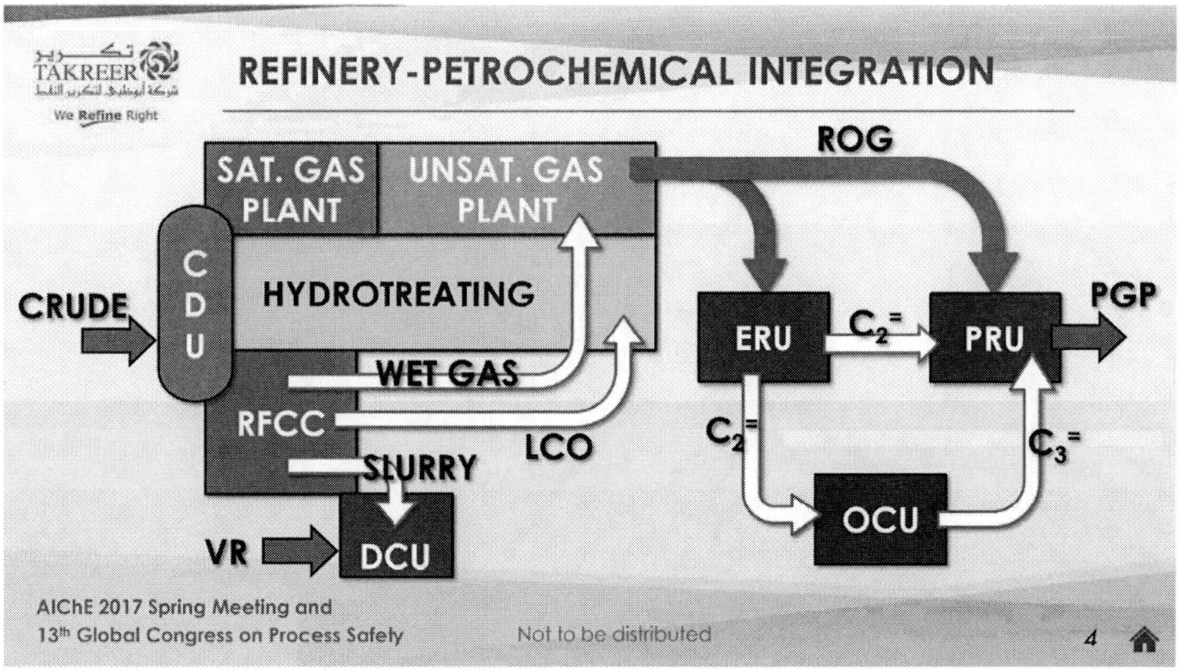
TAKREER
We Refine Right
REFINERY-PETROCHEMICAL INTEGRATION
SAT. GAS PLANT
UNSAT. GAS PLANT
ROG
CDU
CRUDE
HYDROTREATING
WET GAS
RFCC
LCO
SLURRY
VR
DCU
ERU
$C_2^=$
PRU
PGP
$C_2^=$
$C_3^=$
OCU
AIChE 2017 Spring Meeting and
13th Global Congress on Process Safety
Not to be distributed
4

TAKREER
We Refine Right

REFINERY-PETROCHEMICAL INTEGRATION

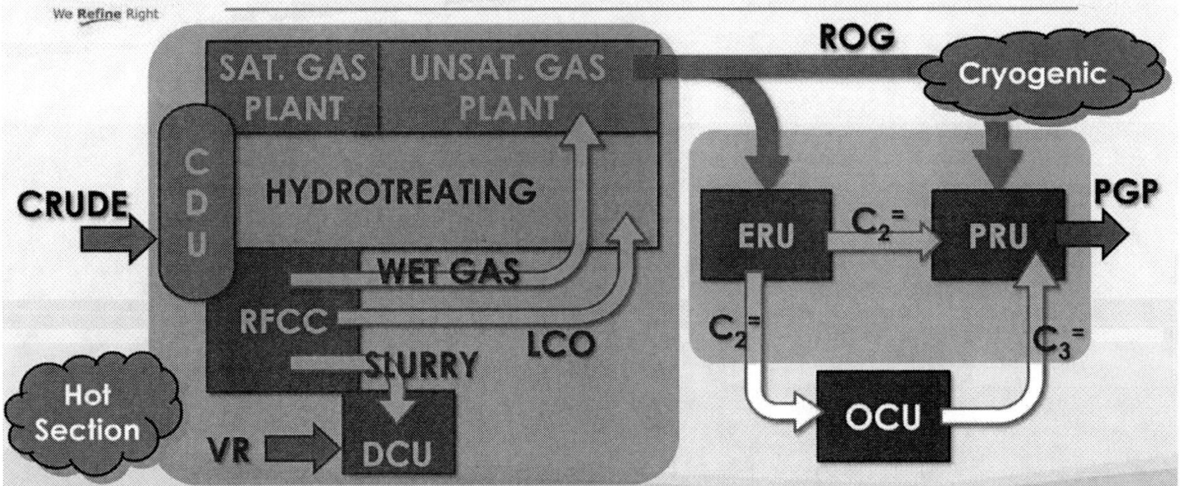

TAKREER
We Refine Right

BACKGROUND OF THE ETHYLENE RECOVERY PROCESS AND THE DEC_2 CIRCUIT

PROCESS BACKGROUND

TAKREER — We Refine Right

Refinery off-gas composition	
Components	**Weight composition %**
Water	0–1
Hydrogen	1.5–2
Nitrogen	15–20
Methane	15–25
Ethylene	20–30
Ethane	10–15
Propylene	10–20
Propane	1.5–2
C_4+	Balance

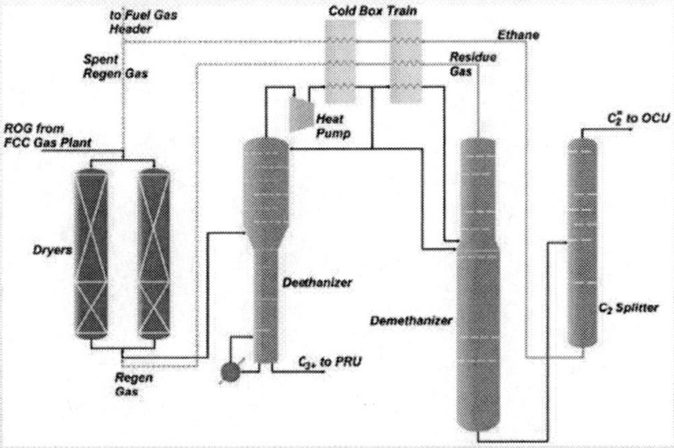

AIChE 2017 Spring Meeting and
13th Global Congress on Process Safety

7

FOCUS ON THE DEC$_2$ CIRCUIT

TAKREER — We Refine Right

Cryogenic Process!

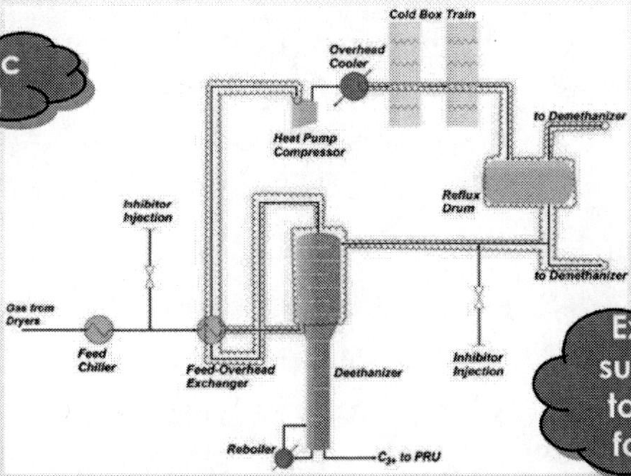

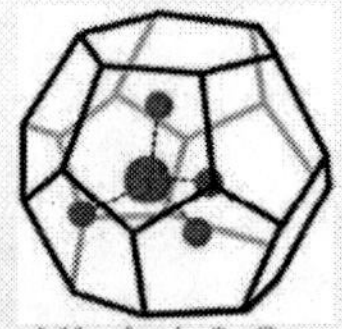

Type I Hydrate*: One CH_4 molecule entrapped in a tetradecahedral water cage

Extremely susceptible to hydrate formation

* Retrieved from Stanford.edu, "**Natural Gas Hydrates**", Sara E. Harrison

AIChE 2017 Spring Meeting and
13th Global Congress on Process Safety

8

TAKREER
We Refine Right

MODEL OF HYDRATE FORMATION AND USE OF SIMULATION FOR ITS PREVENTION

TAKREER
We Refine Right

PROCESS & HYDRATE FORMATION MODEL

- Soave-Redlick Kwong (**SRK**) and Peng-Robinson (**PR**) EoS commonly used for modeling of phase equilibria in the oil & gas industry
- Parameters available for a large number of compounds
- Well-established characterization methods for determination of the parameters
- In case of hydrogen bonding, however, neither of the EoS is capable of accounting for the so-called "**association effect**"
- One newer model is the **SAFT** (Statistical Associating Fluid Theory)
- Or the Cubic Plus Association (**CPA**) – can handle complex systems way simpler than SAFT

PROCESS & HYDRATE FORMATION MODEL

TAKREER
We Refine Right

CPA Model

$$P = \underbrace{\frac{RT}{V_m - b} - \frac{a(T)}{V_m(V_m + b)}}_{\text{SRK contribution}}$$

$$\underbrace{-\frac{1}{2}\frac{RT}{V_m}\left(1+\rho\frac{\partial \ln g}{\partial \rho}\right)\sum_i x_i \sum_i (1 - X_{A_i})}_{\text{Association part from statistical thermodynamics}}$$

$$a(T) = a_0\left[1 + c_1\left(1 - \sqrt{T_R}\right)\right]^2$$

$$X_{A_i} = \frac{1}{1 + \frac{1}{V_m}\sum_j x_j \sum_{B_j} \Delta^{A_i B_j}}$$

$$\Delta^{A_i B_j} = g(V_m)^{\text{ref}}\left[\exp\left(\frac{\varepsilon^{A_i B_j}}{RT}\right) - 1\right] b_{ij} \beta^{A_i B_j}$$

Five parameters (in red) to be estimated

See the following for more details on CPA EoS and mixing rules:
L.C. dos Santos et al., "Cubic Plus Association Equation of State for Flow Assurance Projects", Ind. Eng. Chem. Res., 2015, 54, 6812-6824

PROCESS & HYDRATE FORMATION MODEL

TAKREER
We Refine Right

van der Waals-Platteeuw Model (Water fugacity in hydrate phases)

- Based on $f_w^V(T,P,y) = f_w^L(T,P,x) = f_w^H(T,P)$

$$\Delta\mu_w^H = \mu_w^H - \mu_w^\beta = -RT\sum_m v_m \ln\left(1 - \sum_j \theta_{mj}\right)$$

$$\theta_{mj} = \frac{C_{mj} f_j}{1 + \sum_l C_{ml} f_l}$$

$$C_{mj}(T) = \frac{4\pi}{kT}\int_0^\infty \exp\left(\frac{-w_{mj}(r)}{kT}\right) r^2 dr$$

- μ_w^β : Chemical potential of water in a hypothetical empty lattice
- μ_w^H : Chemical potential of water in a filled cage
- v_m : Number of cavities of type *m* per molecule of water in the lattice
- θ_{mj} : Fraction of type *m* cavities occupied by component *j*
- C_{mj} : Langmuir adsorption constant for component *j* in type *m* cavity
- f_j : Fugacity of component *j*

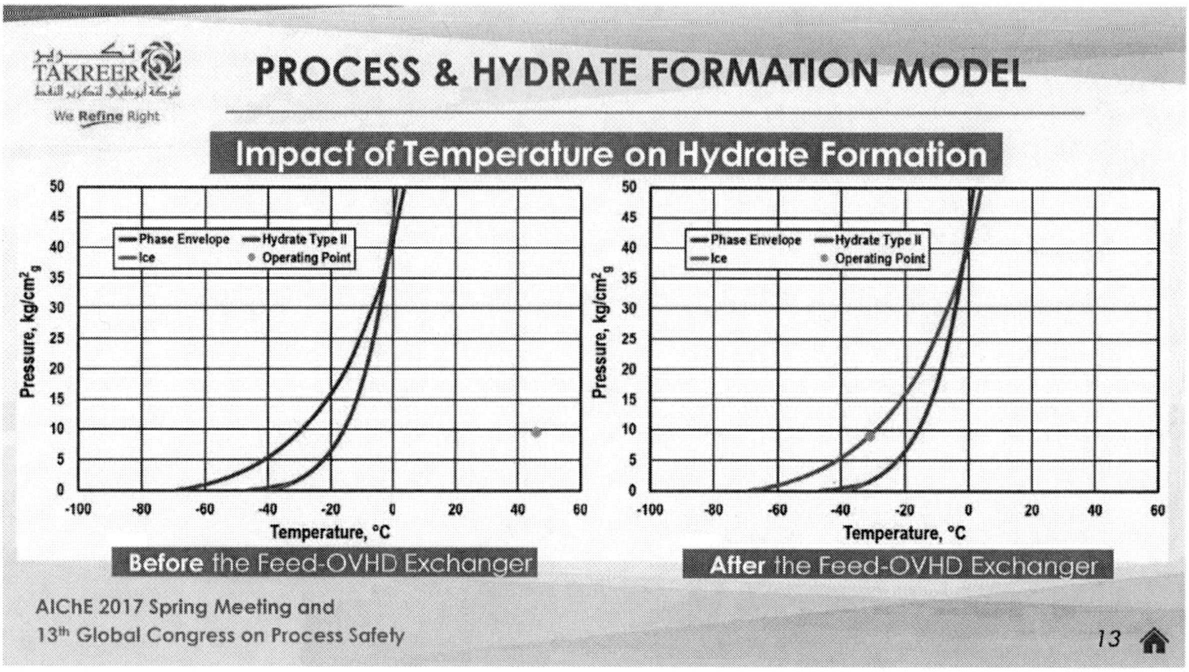
TAKREER
We Refine Right
PROCESS & HYDRATE FORMATION MODEL
Impact of Temperature on Hydrate Formation
Phase Envelope
Hydrate Type II
Ice
Operating Point
Pressure, kg/cm^2_g
Temperature, °C
Before the Feed-OVHD Exchanger
After the Feed-OVHD Exchanger
AIChE 2017 Spring Meeting and
13th Global Congress on Process Safety
13

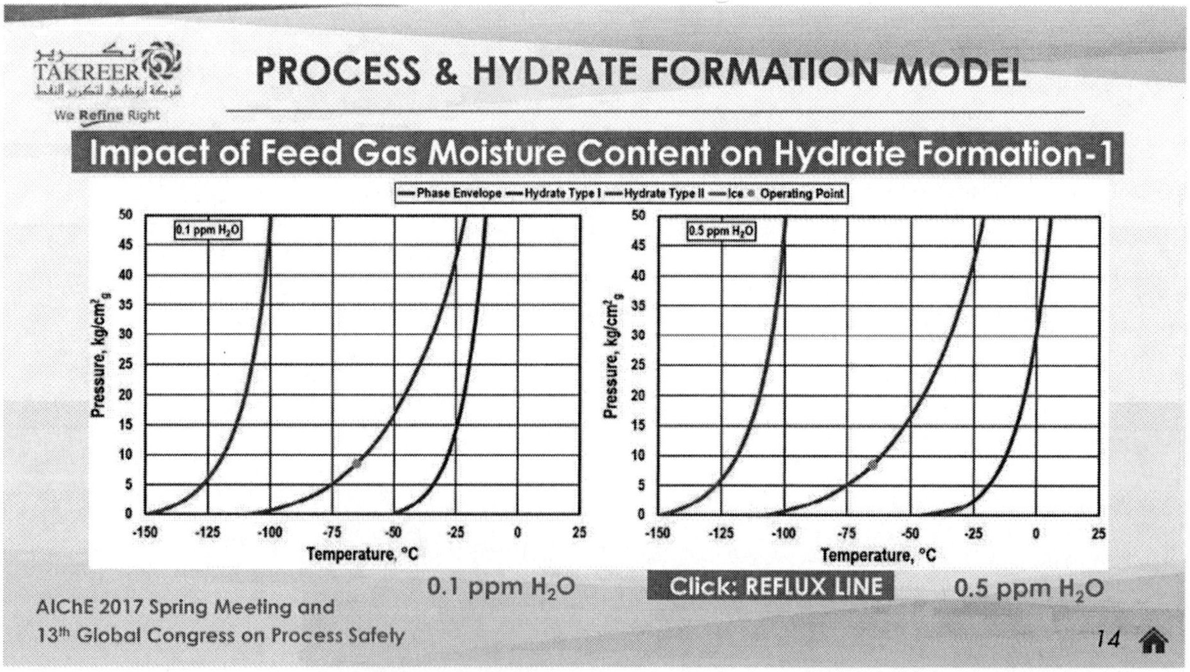
TAKREER
We Refine Right
PROCESS & HYDRATE FORMATION MODEL
Impact of Feed Gas Moisture Content on Hydrate Formation-1
Phase Envelope
Hydrate Type I
Hydrate Type II
Ice
Operating Point
0.1 ppm H_2O
0.5 ppm H_2O
Pressure, kg/cm^2_g
Temperature, °C
0.1 ppm H_2O
Click: REFLUX LINE
0.5 ppm H_2O
AIChE 2017 Spring Meeting and
13th Global Congress on Process Safety
14

TAKREER We Refine Right

PROCESS & HYDRATE FORMATION MODEL

Impact of Feed Gas Moisture Content on Hydrate Formation-2

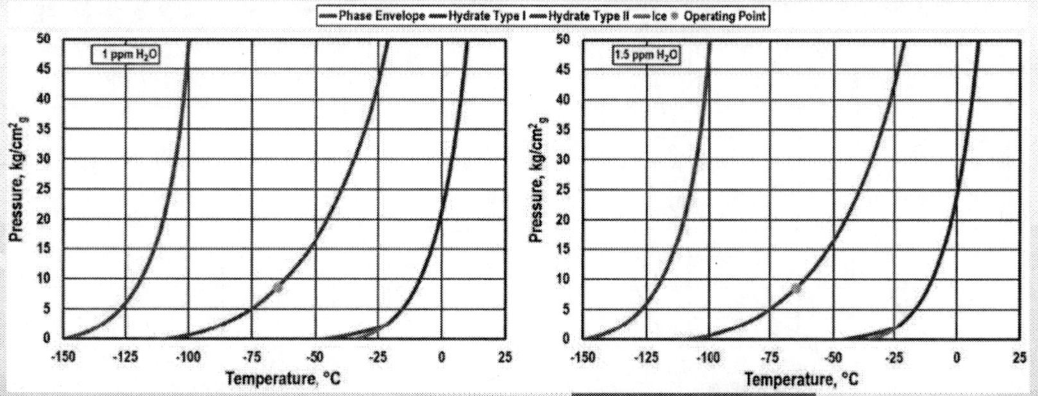

1 ppm H_2O Click: REFLUX LINE 1.5 ppm H_2O

AIChE 2017 Spring Meeting and
13th Global Congress on Process Safety

15

TAKREER We Refine Right

PROCESS & HYDRATE FORMATION MODEL

Hydrate Formation inside the DeC_2 Column-1

for 1 ppm H_2O in the feed

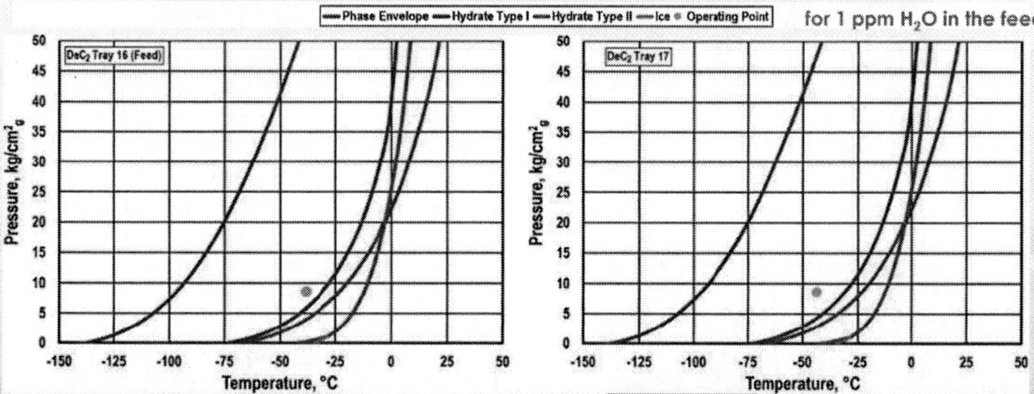

Feed Tray Click: DEC_2 COLUMN 17th Tray

AIChE 2017 Spring Meeting and
13th Global Congress on Process Safety

16

PROCESS & HYDRATE FORMATION MODEL

TAKREER
We Refine Right

Hydrate Formation inside the DeC_2 Column-2

for 1 ppm H_2O in the feed

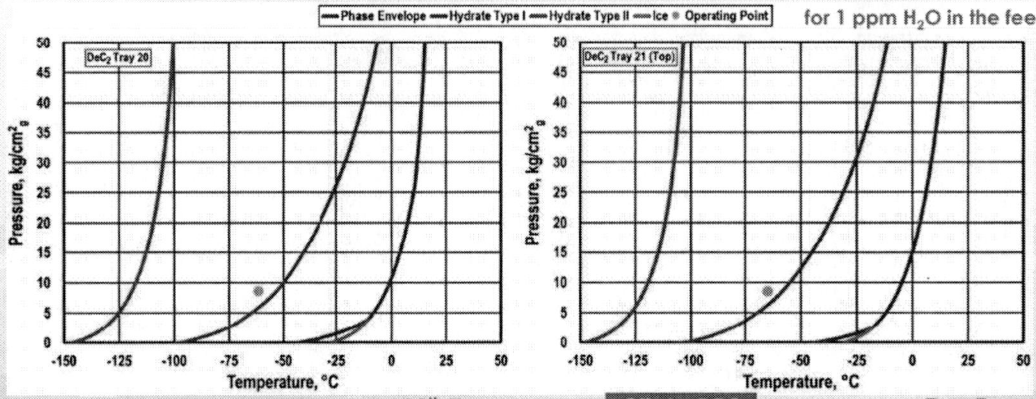

20th Tray

Click: DEC2 COLUMN

Top Tray

PREVENTION OF HYDRATE FORMATION

TAKREER
We Refine Right

- Recent strategies tend to allow hydrates to form but **only to flow** with the fluid
- Agglomeration of the particles prevented by low dosage **kinetic hydrate inhibitor**s
- Traditional approach: Inject **thermodynamic hydrate inhibitor**s to lower the mixture melting point
- Typically lower alcohols (**MeOH**, **EtOH**) and glycols (**MEG**, **DEG**, **TEG**)
- Dry alcohol corrosivity can be problematic in the circuit
- Off-spec product with alcohol contamination
- Hence, guidelines need to be followed in **unit operation and monitoring**!

TAKREER
تكرير
شركة أبوظبي لتكرير النفط
We Refine Right

PREVENTION OF HYDRATE FORMATION

Impact of MeOH Injection for Hydrate Inhibition-1

for 1 ppm H_2O in the feed

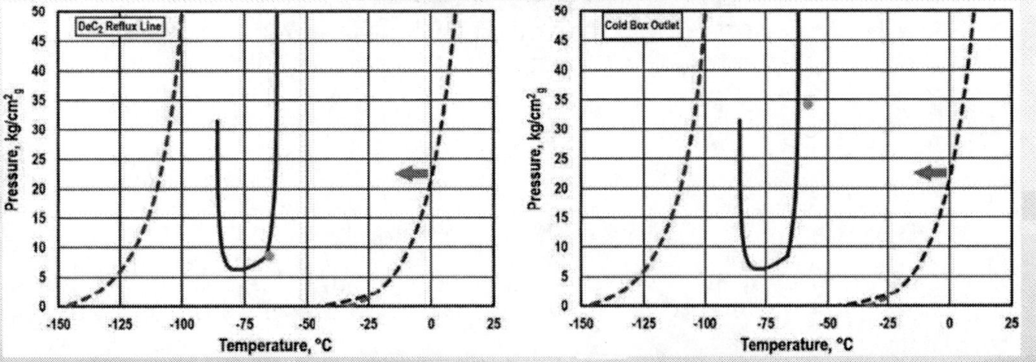

Reflux Line

Click: PFD

Cold Box Outlet

TAKREER
تكرير
شركة أبوظبي لتكرير النفط
We Refine Right

PREVENTION OF HYDRATE FORMATION

Impact of MeOH Injection for Hydrate Inhibition-2

for 1 ppm H_2O in the feed

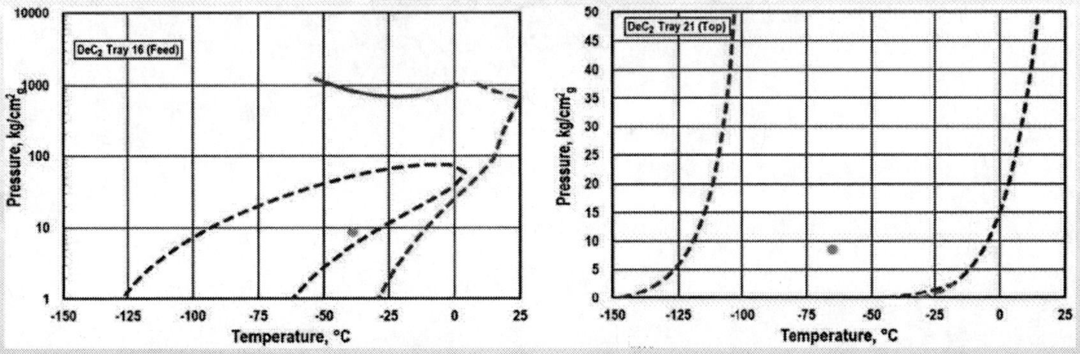

Feed Tray

Click: PFD

Top Tray

TAKREER
We Refine Right

KEY FINDINGS AND CONCLUSIONS

- When wet cracking is followed by cryogenic separation, hydrate formation is highly possible
- Hydrate formation conditions can be proactively determined owing to simulation and engineering tools
- Inhibitors compensate for the upsets occurring upstream (water breakthrough at the gas dryers)
- Prudent practices before shutdown:
  - **Operate the dryers flawlessly and efficiently**
  - **Determine the hydrate forming conditions and the minimum amount of inhibitor required in the worst case scenario of water breakthrough**
  - **Monitor the products for contamination, and flare if required**

AIChE 2017 Spring Meeting and
13th Global Congress on Process Safety

21

تكرير
TAKREER
شركة أبوظبي لتكرير النفط

We **Refine** Right

شركة أبوظبي لتكرير النفط (تكرير)
مركز تكرير للأبحاث
Abu Dhabi Oil Refining Company (TAKREER)
TAKREER Research Centre (TRC)

AIChE 2017 Spring Meeting and
13th Global Congress on Process Safety

SLIDE #4

- Flexible fluid catalytic cracking units (FCCU) – capable of processing a wide range of feeds – are candidates for producing the feedstock for polymer-grade propylene (PGP)
- **ROG**: Refinery off-gas
- **ERU**: Ethylene ($C_2^=$) recovery unit
- **PRU**: Propylene ($C_3^=$) recovery unit
- **OCU**: Olefins conversion unit

SLIDE #5

- Cryogenic processes downstream of the FCC, with temperatures down to -100°C and high pressures
- Pipelines and equipment prone to hydrates, even at low moisture concentrations
- Formation and plugging of hydrates are threats to production and plant integrity
- Ideally gas dryers are installed upstream of the cryogenic section for moisture removal
- Nevertheless, process needs to be analyzed for risks under different scenarios

SLIDE #7

- Refer to table for the typical refinery off-gas composition downstream of the FCC gas plant
- ROG must be dried using fixed beds of mol sieves (one in service, other under regen)
- Moisture analyzer in between the beds essential
- Deethanizer (DeC_2) tower separates C_2 and C_3+ components
- Bottoms with 80 – 85 wt% $C_3^=$ sent to PRU; overheads heat-exchange with the feed and compressed in a heat pump compressor
- Compressed overheads chilled against refrigerants and partially condensed in a train of cold boxes
- Part of the liquid is refluxed to the tower, the remaining is sent to the demethanizer (DeC_1)

SLIDE #7 – CONT'D

- Remaining vapor sent to the heat recovery system (HRS) and to DeC_1
- DeC_1 overheads makes up the residue gas which is used in the HRS and cold boxes
- Dryer beds are also regenerated using the residue gas
- Spent residue gas exported to the fuel gas header
- DeC_1 bottoms sent to the C_2 splitter
- C_2 splitter bottoms (essentially ethane) first used to cool the DeC_2 overheads, and then sent to fuel gas header
- High-purity $C_2^=$ exported to OCU (metathesis reactor between C_2's and C_4's to produce C_3's)
- Serious risk of hydrate formation if water, even in low quantities, is present in the streams

TAKREER
We Refine Right

SLIDE #8

- Issues in the ERU due to upsets in the DeC_2 tower and the overhead circuit leading to plant downtime and flaring of valuable products
- Excessive pressure drop across the tower indicative of hydrate plugging
- Hydrate formation in the reflux drum and line also encountered
- Better understanding of the problem and proper troubleshooting required a simulation model to be built
- Determine the thermodynamic hydrate envelope and effect of water content in the feed gas
- Locations in the circuit that are prone to hydrate formation are marked in red

TAKREER
We Refine Right

SLIDE #10

- Every piece of equipment and stream included in the simulation, but the cold boxes modeled as heaters
- Accurate prediction of the phase behavior under hydrate-forming conditions calls for an equation-of-state (EoS) that can model mixtures of hydrocarbons, water and alcohols
- One EoS that is capable of predicting polar interactions and hydrogen bonding is the Cubic-Plus-Association (CPA)
- CPA is an extension of the cubic Soave-Redlich-Kwong (SRK) EoS that also incorporates the associating terms representing the effect of H-bonding
- CPA is accurate in predicting the partitioning of alcohol and glycol inhibitors between the HC liquid, aqueous liquid and vapor phases

SLIDE #12

- Solid (ice and hydrate) phase behavior predicted using the van der Waals-Platteeuw model
- Equilibrium based on the equality of chemical potential or fugacity of water between phases
- Due to the specific structure of the hydrate, gas molecules are trapped inside cages of H-bonded water molecules
- Usually empirical correlations are used in simulation tools to calculate C_{mj}
- DeC_2 tower modeled as a reboiled absorber
- Number of theoretical stages is 21 with feed location at stage 16
- Top temperature and reboiler boil-up ration chosen as the tower specs

SLIDE #13

- Figures show the impact of temperature on hydrate formation
- While the DeC_2 charge before the feed-overhead exchanger is away from the hydrate-forming region due to warmer temperatures (LHS figure), downstream of the exchanger, temperature falls within the hydrate-forming region that is shown by the red curve (RHS figure)
- Since composition across the HEX remains the same, temperature drop is the only factor that triggers hydrate formation

TAKREER
We Refine Right

SLIDES #14-15

- Impact of water content on the hydrate formation temperature for the reflux line shown
- As the stream water content increases, the hydrate-forming region that pertains to Type I hydrates shift to warmer temperatures (increasing abscissa in the plots)
- Shift is drastic in going from 0.1 to 0.5 ppm, which amounts to a temperature change of almost 15°C at 15 kg. (Slide 14)
- Increase from 0.5 to 1.5 ppm does not have as big an impact (Slide 15)

TAKREER
We Refine Right

SLIDES #16-17

- The hydrate-forming regions from the feed location (Tray 16) of the DeC_2 tower to its top are shown to illustrate the effect of stream or tray composition on hydrate formation
- While the thermodynamic possibility of hydrate formation exists on all trays, Type II hydrates (red curve) will form in Trays 16 and 17 (Slide 16)
- Type II hydrates will form on upper trays (Slide 17)
- Type I hydrates encase small-diameter molecules such as methane or ethane, whereas Type II hydrates accommodate larger guest molecules, typically propane or iso-butane
- The latter heavy molecules are sent to the bottoms product, which make the rectifying section trays prone to Type I hydrates only

SLIDE #18

- Self explanatory...

SLIDES #19-20

- Specific inhibitor injection points located around the DeC_2 tower, one immediately after the feed chiller and one on the reflux line
- Assumption of feed carrying 1 ppm water, minimum combined (at injection points 1+2) injection of MeOH calculated to be 0.3 wt% of the feed required to suppress hydrate formation
- Temperature shift of almost 60°C for the Type I hydrate curves pertaining to the reflux line and cold box outlet is noteworthy (Slide 19)
- Moreover, Type II hydrate curves cease to exist upon MeOH injection (absence of solid red curves)
- Risk of hydrate formation around the vicinity of the operating point completely eliminated, as seen by the absence of solid hydrate curves (Slide 20)

SLIDES #19-20 – CONT'D

- Although inhibition by MeOH solutions technically feasible, this application is limited to -98°C, the freezing point of MeOH
- Use of MEG (monoethylene glycol) alone cannot be considered, as its freezing point is -13°C
- Eutectic mixtures will yield inhibitors with freezing points lower than those of the individual components
- Due to its high relative volatility, MeOH is lost into the vapor product. However, its cost is still very low compared with the investment necessary to recover it from products for reuse (assuming products are not off-spec!)

SLIDE #21

- Self explanatory…

2017 AIChE®
SPRING MEETING&
13TH
GLOBAL CONGRESS
ON PROCESS SAFETY

March 26–30, 2017
Henry B. Gonzalez
Convention Center
San Antonio, TX

17AIChE
Spring Meeting, San Antonio, TX

13TH GLOBAL CONGRESS ON PROCESS SAFETY

Consider the Alternatives - NGL Fractionation Configurations

Scott Schroeder, Advisian
scott.schroeder@advisian.com

www.advisian.com

Advisian
WorleyParsons Group

United States NGL Production

- The development of shale oil and gas reserves in the United States has lead to the production of NGL doubling in the last 10 years.

- This increase in production has led to a downward trend in NGL prices.

United States NGL Production

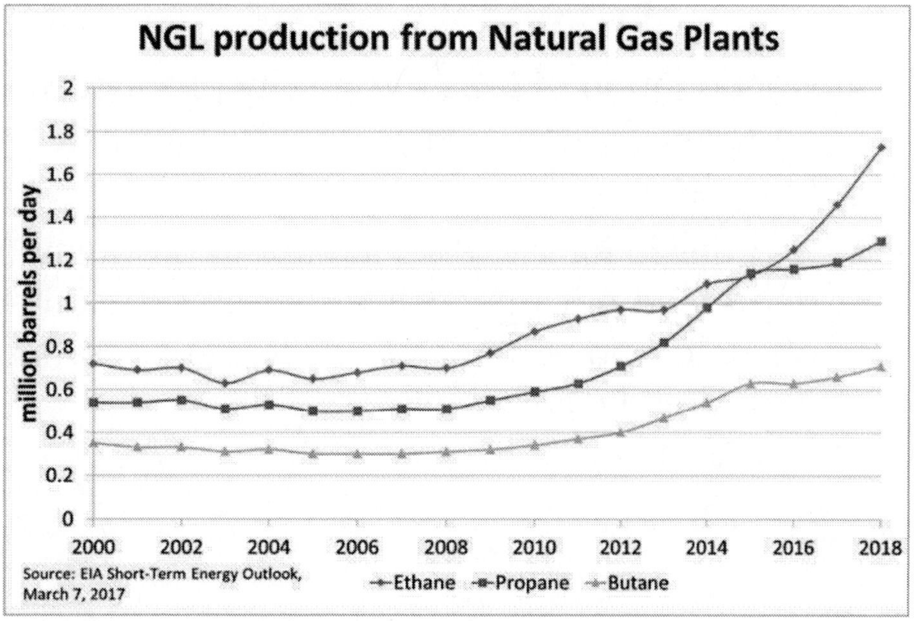

Propane Price

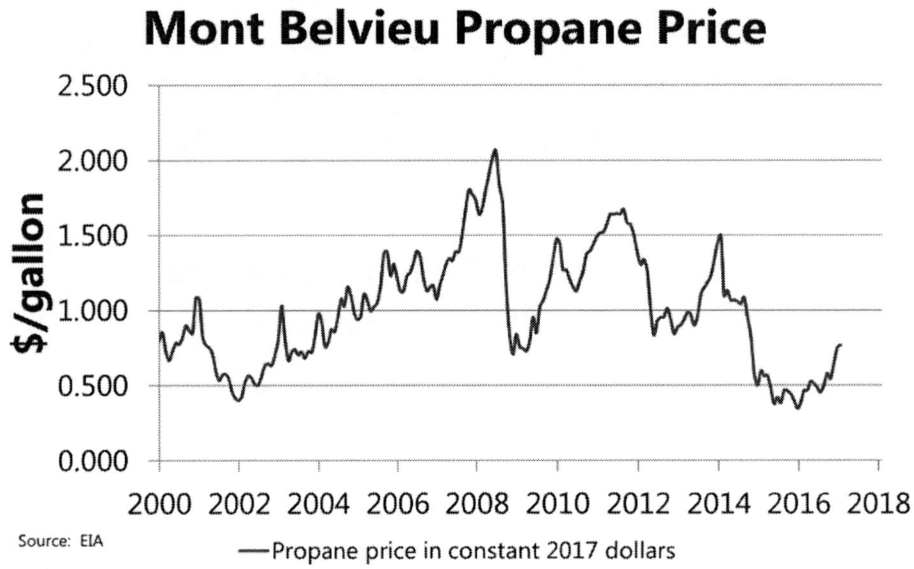

United States NGL Production

- Fractionation of NGL into purity products is performed using energy intensive distillation operations.
- In order to handle the increase in production at depressed pricing levels attention should be given to opportunities to decrease operating and capital costs.

NGL Fractionation Sequence

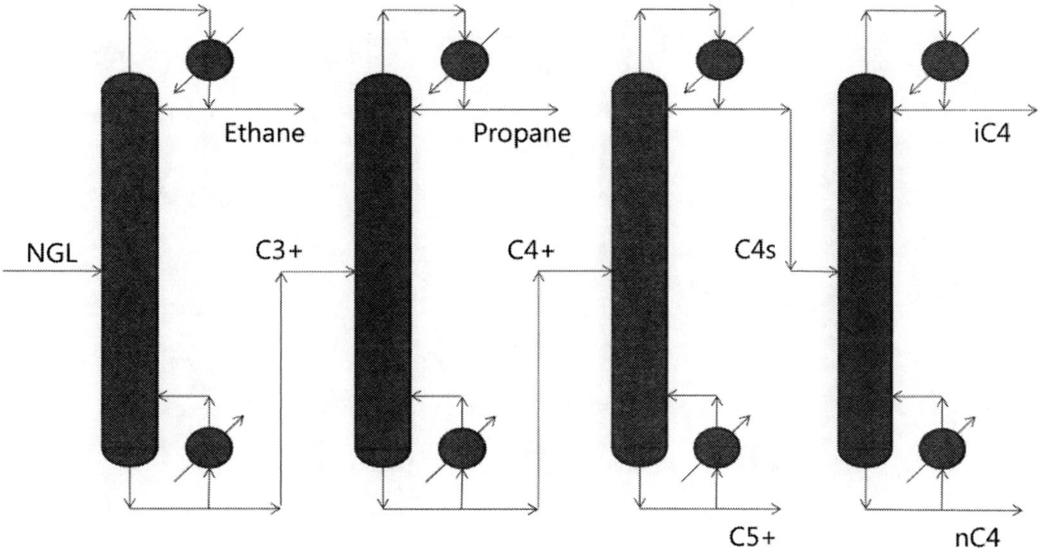

Simple Heat Recovery

Addition of heat exchangers to optimize energy usage between product and feed streams.

- Low capital cost
- Well understood by operations staff
- Limited opportunity – most new or operating facilities will have heat recovery included in the design

Simple Heat Recovery

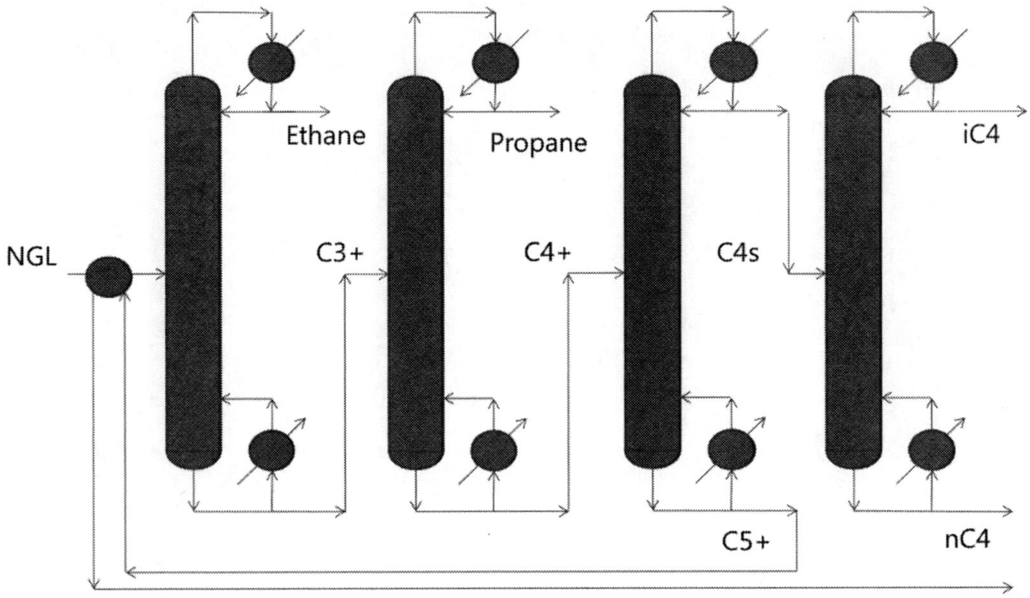

Thermally Coupled Columns and Divided Wall Columns

- Conventional column sequences for distilling multicomponent mixtures are well known to design engineers and operations staff. There are well established design and operations guidelines resulting in units that are easy to control and operate.

- However, in each column there is an inherent inefficiency as components are remixed. Interconnecting the columns can reduce this inefficiency while reducing the number of condensers and reboilers. This can result in considerable savings in capital and operating costs.

Conventional Column Sequences

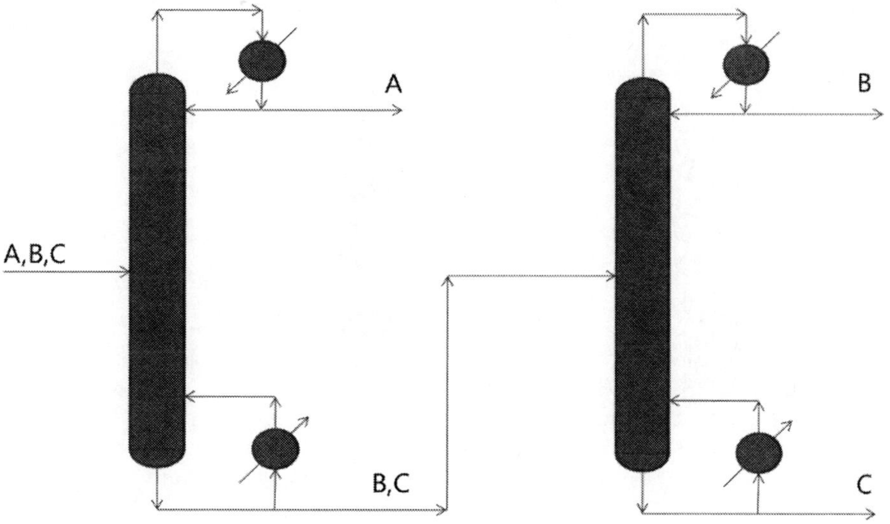

2017 AIChE® SPRING MEETING & 13TH GLOBAL CONGRESS ON PROCESS SAFETY

Advisian / 11

Conventional Column Sequences

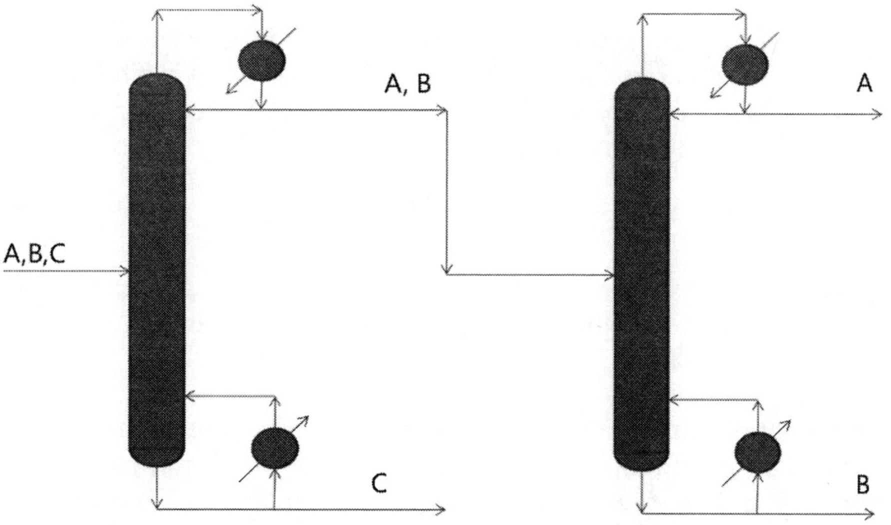

2017 AIChE® SPRING MEETING & 13TH GLOBAL CONGRESS ON PROCESS SAFETY

Advisian / 12

Petlyuk Column

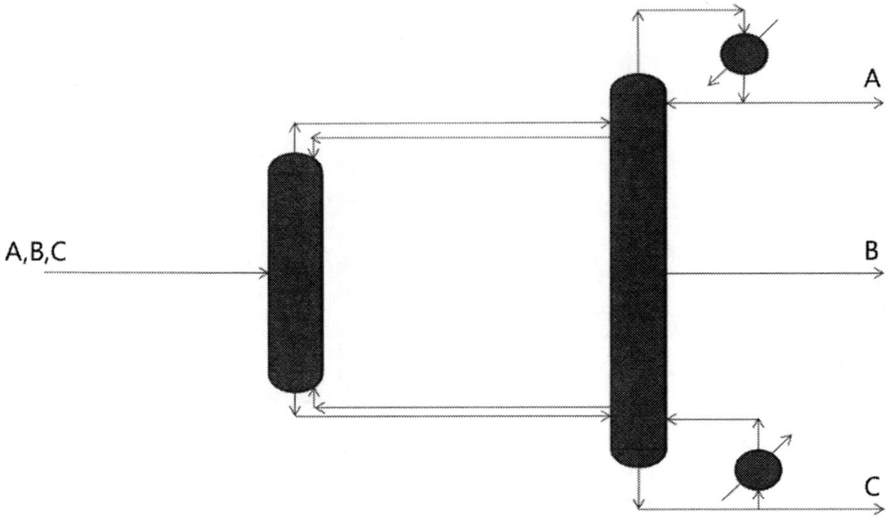

Thermally Coupled Columns and Divided Wall Columns

- Divided Wall Columns combine two or more thermally coupled columns into a single column by installing a partition wall in the column. Assuming minimal heat transfer across the partition wall they are thermodynamically equivalent to thermally coupled columns.

- This can result in further capital savings because of the reduction in number of installed columns and interconnecting piping.

- One limitation of the divided wall columns is that entire column operates at one pressure. Combining sequences that normally operate at different pressure can result in an unfavourable shift in utility conditions.

Divided Wall Column (DWC)

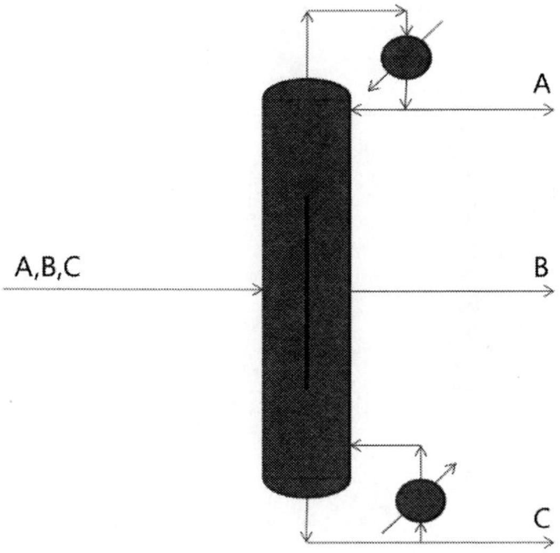

2017 AIChE® SPRING MEETING & 13TH GLOBAL CONGRESS ON PROCESS SAFETY

Advisian / 15

Thermally Coupled Column

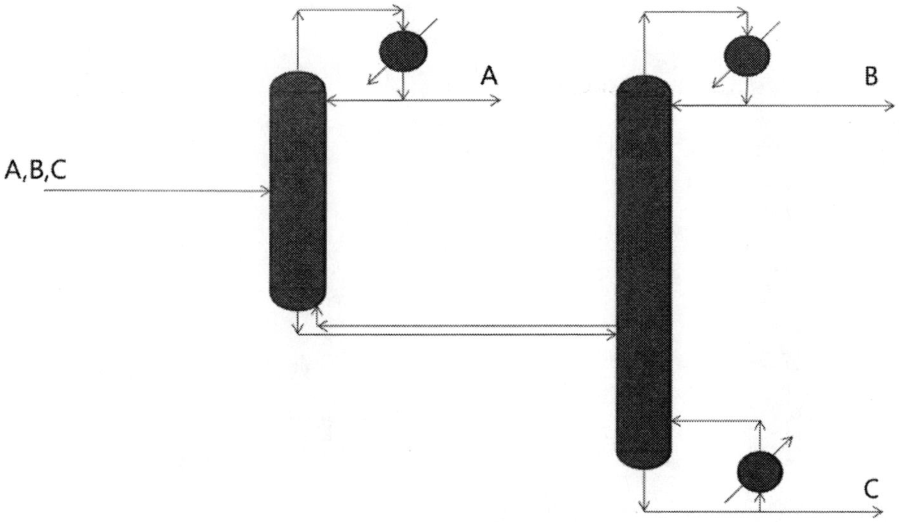

2017 AIChE® SPRING MEETING & 13TH GLOBAL CONGRESS ON PROCESS SAFETY

Advisian / 16

Divided Wall Column (DWC)

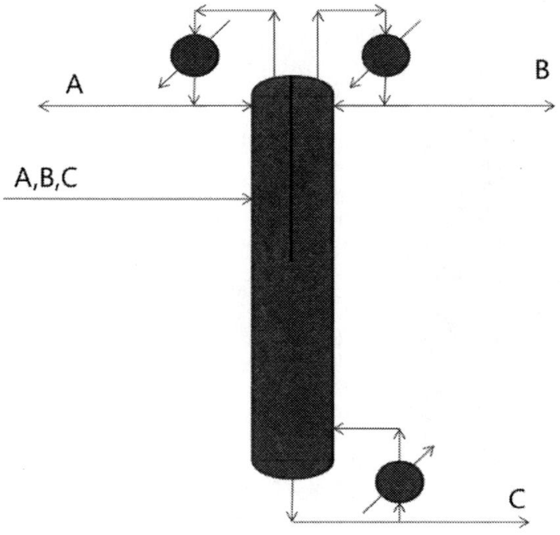

2017 AIChE® SPRING MEETING & 13TH GLOBAL CONGRESS ON PROCESS SAFETY

Advisian / 17

Thermally Coupled Column

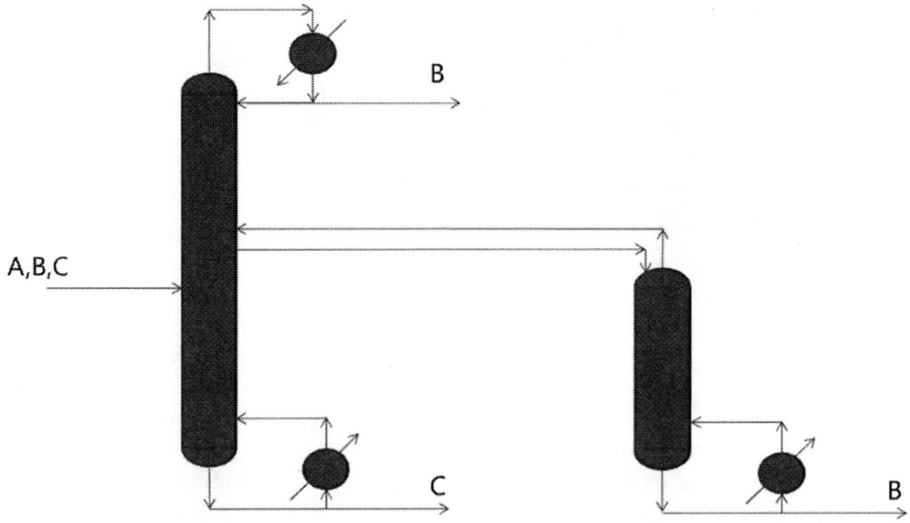

2017 AIChE® SPRING MEETING & 13TH GLOBAL CONGRESS ON PROCESS SAFETY

Advisian / 18

Divided Wall Column (DWC)

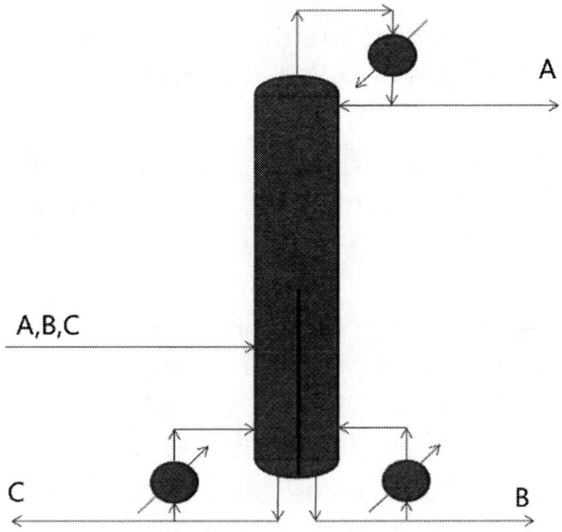

Divided Wall Column (DWC) – 4 component

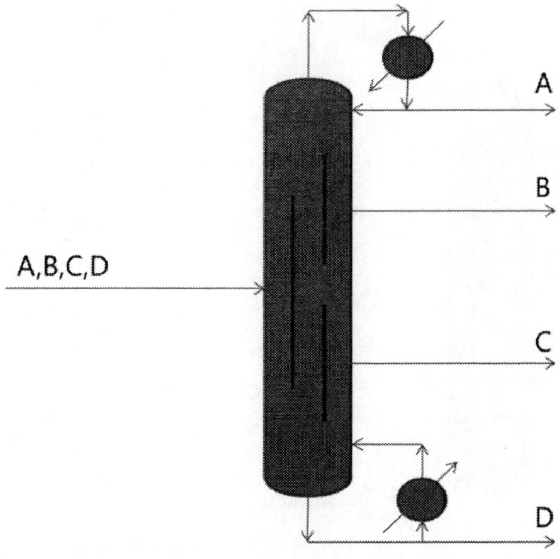

Heat Pumps

- Distillation involves adding heat into the reboiler at the bottom of the column, most of which is removed at the condenser at the top of the column.
- The heat for the reboiler is generally generated by the combustion of a hydrocarbon fuel. The heat removed in the condenser is generally rejected into the atmosphere. This is a relatively inefficient use of energy.
- However the heat removed from the condenser cannot be directly applied to the reboiler because of the temperature difference. A heat pump creates a thermodynamic cycle to upgrade the heat from the condenser so it can be used in the reboiler.
- Heat pumps are most useful when the temperature difference between the condenser and reboiler is small and the duty is large.

Heat Pump

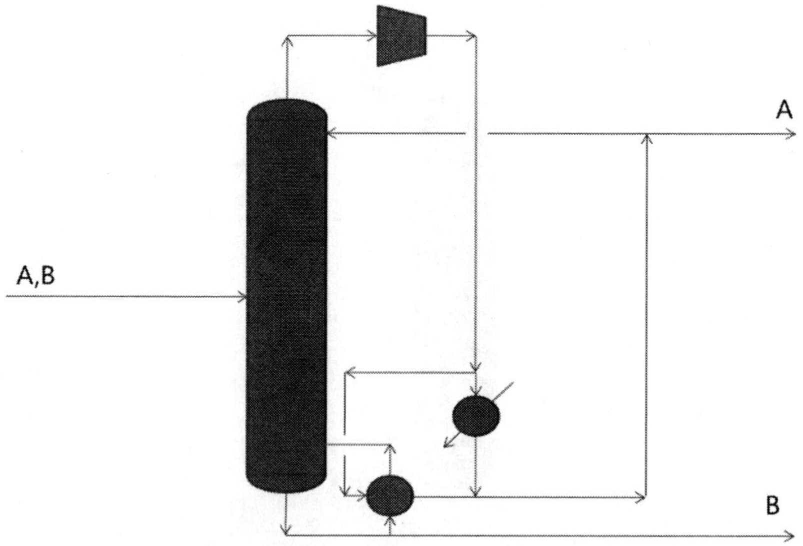

Heat Pump

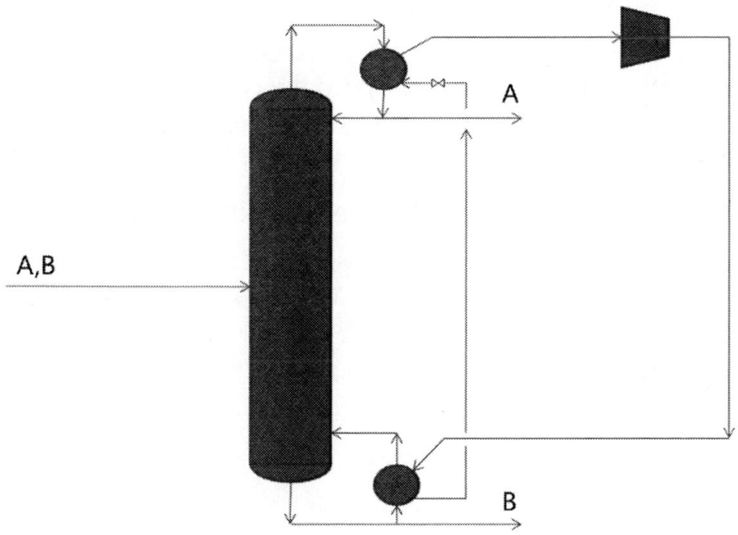

Heat Pumps

Although heat pumps can greatly decrease the fired heat duty required they have several disadvantages -

- Capital Cost – addition of a compressor increases cost
- Maintenance – rotating equipment increase maintenance cost and decreases overall availability factor
- Project Schedule – a custom compressor for maximum efficiency will likely become the critical path in the project schedule

If air emissions are a critical issue, then the value of a heat pump to reduce fired heat duty can be more important than the operating cost saved by decreased fuel usage.

NGL Fractionator Configuration

- Beyond the simple heat recovery between feeds and products a combination of heat pumps, thermally coupled columns and divided wall columns can be used to create and overall more efficient and cost effective NGL Fractionation Train.
- These process options can be used to debottleneck existing facilities
- The optimum configuration depends on local factors -
  - Feed Composition
  - Cost of Utilities – electricity vs. fuel
  - Fabrication and Construction Cost
  - Regulatory Issues

NGL Fractionator Configuration

General Trends

- Heat exchange should be consider for any high or low temperature product streams
- Limits on fired emissions will likely drive the design to use heat pumps or other more complicated thermal integration strategies.
- Thermally coupled columns and divided wall columns are not new - they date back the 1940's – however they are they have not been used widely and will typical be unfamiliar to operations staff.
- Divided Wall Columns (DWC) have been gaining in attention for the combined energy and capital cost savings that are possible. There are now over 100 in operation.
- DWC are not the correct solution for every problem. NGL fractionation sequences typically operate at different pressures. The advantages of the DWC must be weight against the penalty in operating pressures.

Questions

??

Advisian

WorleyParsons Group

17 AIChE Spring Meeting, San Antonio, TX

13TH GLOBAL CONGRESS ON PROCESS SAFETY

2017 AIChE Spring Meeting
March 28, 2017
Henry B. Gonzalez Convention Center
San Antonio, TX

Development of Fuel Processor for Residential PEFC System

Tokyo Gas CO., LTD
Masafumi Ohashi, Yoshitaka Baba, Shin Inagaki

TOKYO GAS 1

Executive Summary (1/2)

Introduction

- ✓ In 2003, Tokyo Gas developed the world's highest level of efficiency and compactness fuel processor(FP) for residential PEFC systems. At the same time, this component can be mass produced at low cost.
- ✓ However, for the PEFC systems to become more widespread, further cost reduction of the FP is required.

⇒ We try to reduce the amount of catalyst in order to reduce FP cost.

Approach to reducing the amount of catalyst

- ✓ CO is generated from city gas in the presence of Steam Reforming(SR) catalyst.
- ✓ PEFC electrodes are poisoned by CO.
- ✓ CO is removed by a water gas shift (WGS) catalyst.
- ✓ Because of the large amount of WGS catalyst, the structure of FP was complex.

The amount of WGC catalyst is determined by CO concentration at SR catalyst.

⇒ We try to reduce the CO concentration of SR catalyst to bring down the amount of WGS catalyst.

Outlet temperature of SR catalyst → Lower ➡ CO concentration → Decrease; Amount of H_2 → Decrease

$S/C(H_2O)$ → Increase ➡ Amount of H_2 → Increase

$S/C(H_2O)$ → Increase = High S/C
Outlet temperature of SR catalyst → Lower

We designed the high-S/C-compliant FP...

TOKYO GAS 2

Executive Summary (2/2)

Structure of the high-S/C-compliant FP

- ✓ To reduce the amount of WGS catalyst, the structure is simplified.
- ✓ The flame is positioned at the same height as the present FP.

Although the amount of water increased, vaporization capability is the same as the present FP, by manufacturing the vaporizer using a new method,

New method of vaporizer manufacturing

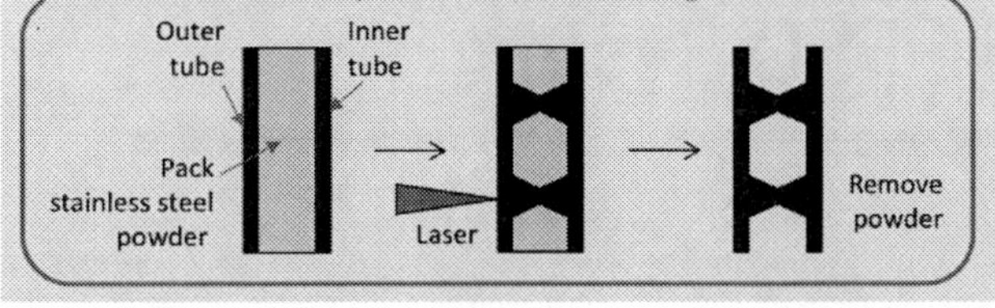

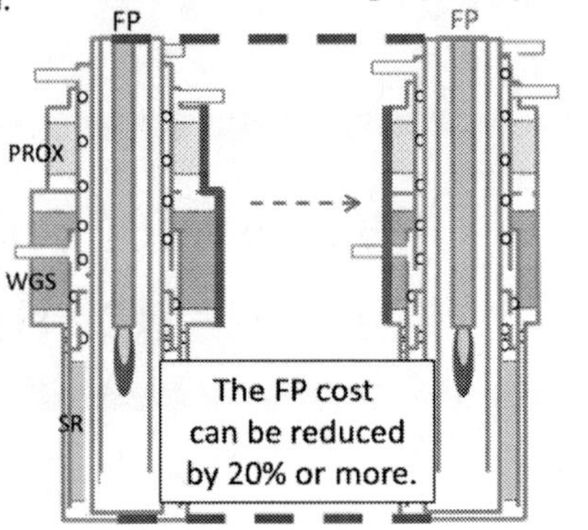

The results of process simulation and FP performances test

- ✓ Process simulation was conducted using the VMG Sim^{TM} (Virtual Materials Group, Inc).
- ✓ In the FP performance test, each fluid were supplied by controlling the flow rate with the mass flow controller.
  The reformed gas was analyzed with a gas chromatograph (GC/TCD,Shimadzu, GC-8A).

⇒ The performance of the high-S/C-compliant FP corresponded to the performance of the present FP and the calculation result.

The FP cost could be reduced by 20% or more.

TOKYO GAS 3

Outline

- **Introduction**
  - System structure of residential PEFC system
  - Fuel Processor (FP)
  - History of FP development in Tokyo Gas
  - Objective
- **Process simulation**
  - Approach to reducing the amount of WGS cat.
  - Process simulation method
  - Calculation result
- **FP performances test**
  - Structure of the high-S/C FP and the advanced high-S/C FP
  - FP performance testing equipment
  - Performance of the high-S/C FP and the advanced high-S/C FP
- **Summary**

TOKYO GAS 4

System structure of residential PEFC system

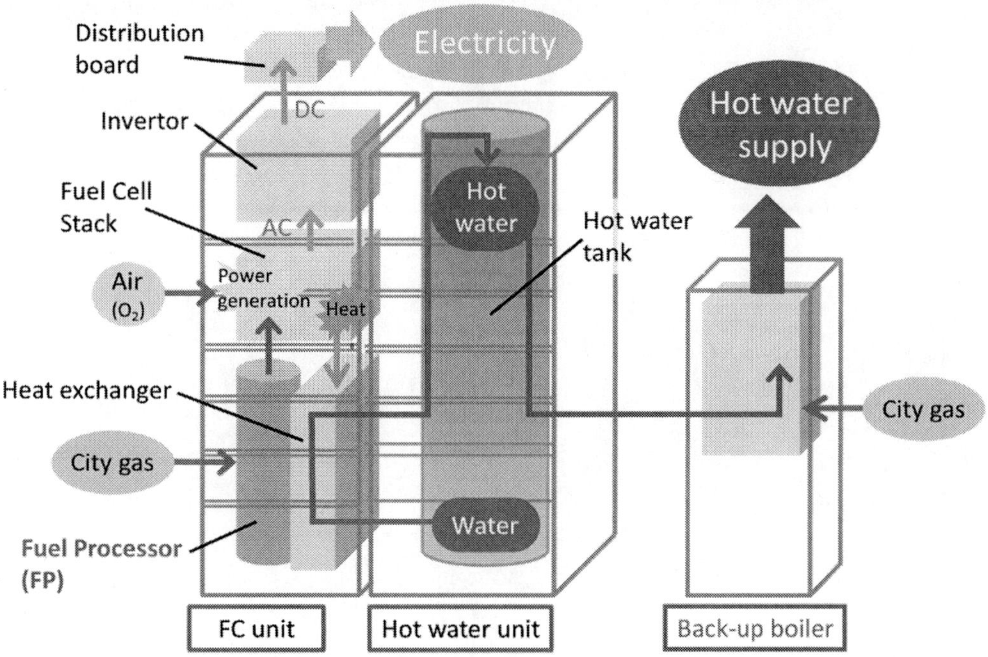

Fuel Processor(FP)

Fuel Processor(FP) : A system to produce hydrogen from city gas

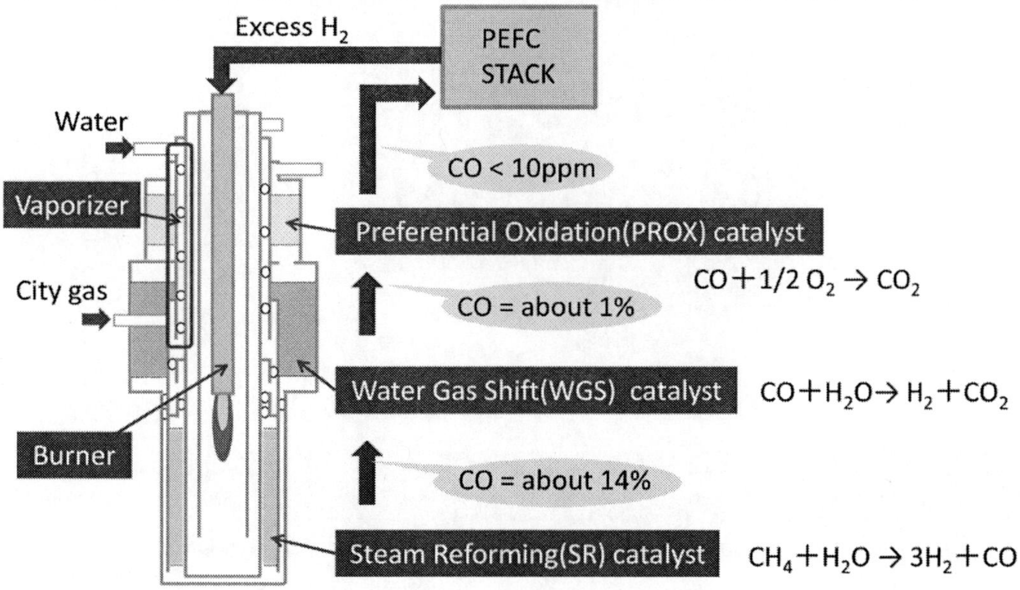

History of FP development in Tokyo Gas

Tokyo Gas embarked on the development of FP

Tokyo Gas succeeded commercialization of ENE-FARM.

New model of ENE-FARM released

1998 — 2000 — 2003 — 2009 — 2015

Tokyo Gas developed an integrated FP

New model released

Production costs were reduced by half.
Volume was also reduced by about 2/3.

TOKYO GAS 7

History of FP development in Tokyo Gas

Tokyo Gas embarked on development of FP

New model of ENE-FARM released

2000 2003 1998 2015

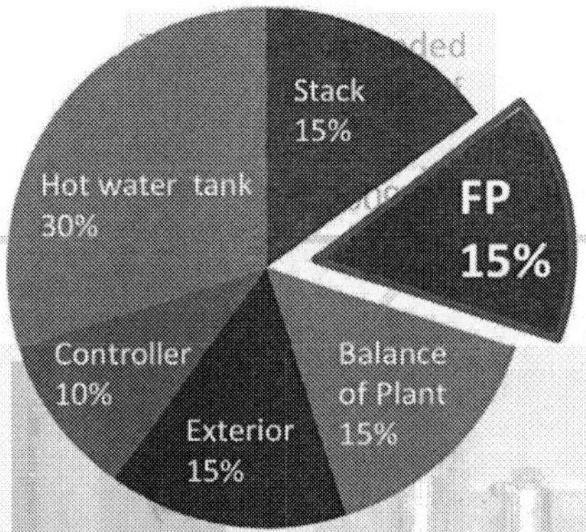

Cost structure of residential PEFC system (METI,2014/2)

Tokyo Gas developed an in

by about 2/3.

The further reduction of FP cost is necessary, for the widespread of residential PEFC system.

TOKYO GAS 8

Objective

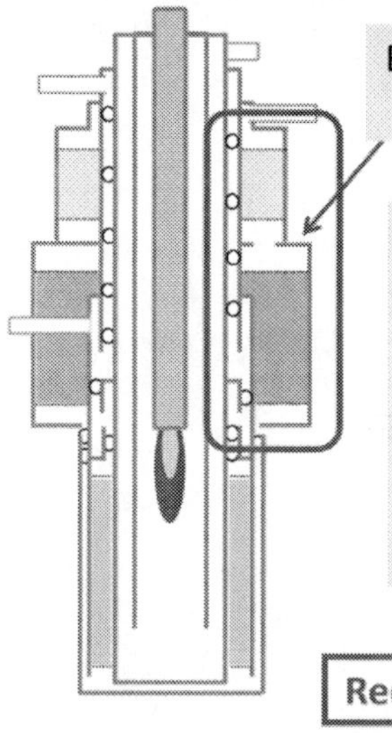

Because of the large amount of WGS catalyst, the structure was complex.

To reduce the amount of WGS catalyst, the structure is simplified.

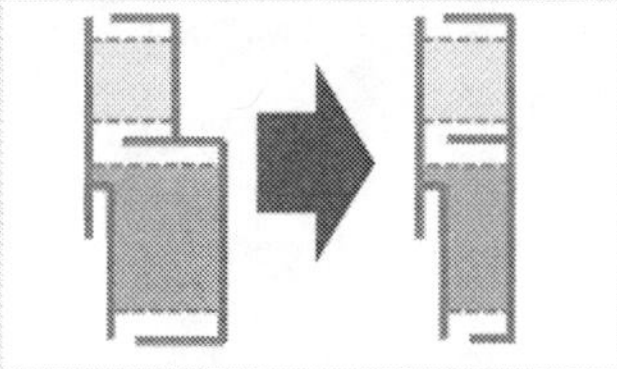

Reducing the production costs and the catalyst cost

We try to reduce the amount of WGS catalyst in order to reduce FP cost.

Approach to reducing the amount of WGS cat.

Performance of the present FP

Operating condition	
Steam/Carbon ratio (S/C)	2.5
Outlet Temp. of SR catalyst	680°C
Performances	
H_2 generation efficiency [*1]	84%HHV
H_2 utilization ratio[*2]	73%
CH_4 conversion	92%
CO concentration at SR catalyst	14%
CO concentration at PROX catalyst	<10ppm

[*1] H_2 generation efficiency= $\frac{Q_R - Q_B}{Q_{input}}$

[*2] H_2 utilization ratio = $\frac{Q_R - Q_B}{Q_{H2}}$

Q_R : calorific value of Reforming gas
Q_B: calorific value of fuel for burner
Q_{H2}: calorific value of H_2 in Reforming gas

We try to reduce the CO concentration of SR catalyst to bring down the amount of WGS catalyst.

Approach to reducing the amount of WGS cat.

Steam reforming reaction
$CH_4 + H_2O = 3H_2 + CO - 206$ kJ

Water gas shift reaction
$CO + H_2O = H_2 + CO_2 + 42$ kJ

Outlet temperature of SR catalyst → Lower ➡ CO concentration → Decrease; Amount of H_2 → Decrease

S/C(H_2O) → Increase ➡ Amount of H_2 → Increase

S/C(H_2O) → Increase
Outlet temperature of SR catalyst → Lower

Process simulation method

【Process simulator】VMG Sim™

【Parameter】・S/C ・Outlet temperature of SR catalyst

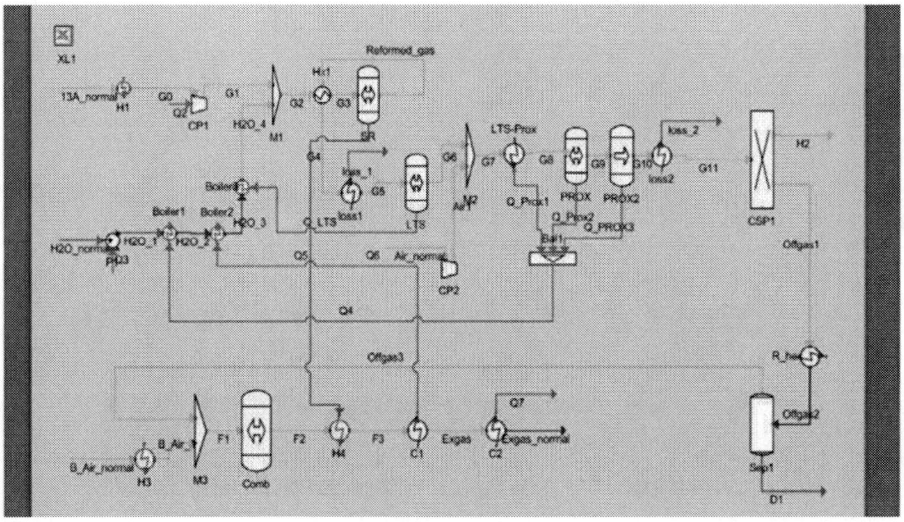

Calculation result

		Present	Calculations		
			#1	#2	#3
Condition	S/C	2.5	4	4.5	5
	Outlet temp. of SR cat.	680°C	580°C	565°C	555°C
Result	CH_4 conversion	92%	76%	75%	75%
	CO concentration at SR catalyst	14%	5.3%	4.3%	3.6%
	H_2 generation efficiency	84%	80%	79%	80%
	H_2 utilization ratio	73%	85%	85%	85%
	CO concentration at PROX catalyst	<10ppm	-	-	-

◎ CO concentration of SR catalyst can be decreased by about 60%.
△ H_2 generation efficiency was decreased.
△ H_2 utilization ratio was increased.

Structure of high-S/C FP

Present FP

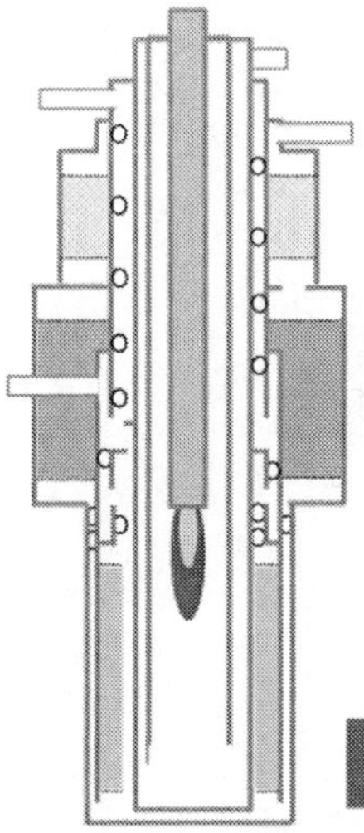

(1)
The amount of WGS catalyst was reduced by half.

(2)
The flame was repositioned to a higher location.

(3)
The structure was simplified.

High-S/C FP

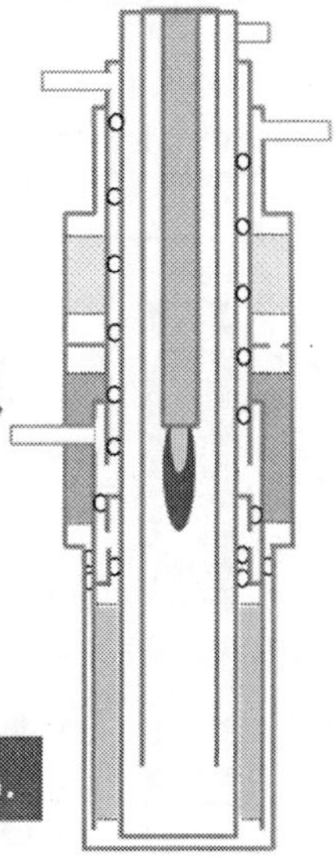

The FP cost can be reduced by 20%.

But... the area of vaporizer was increased.

Concept of advanced high-S/C FP

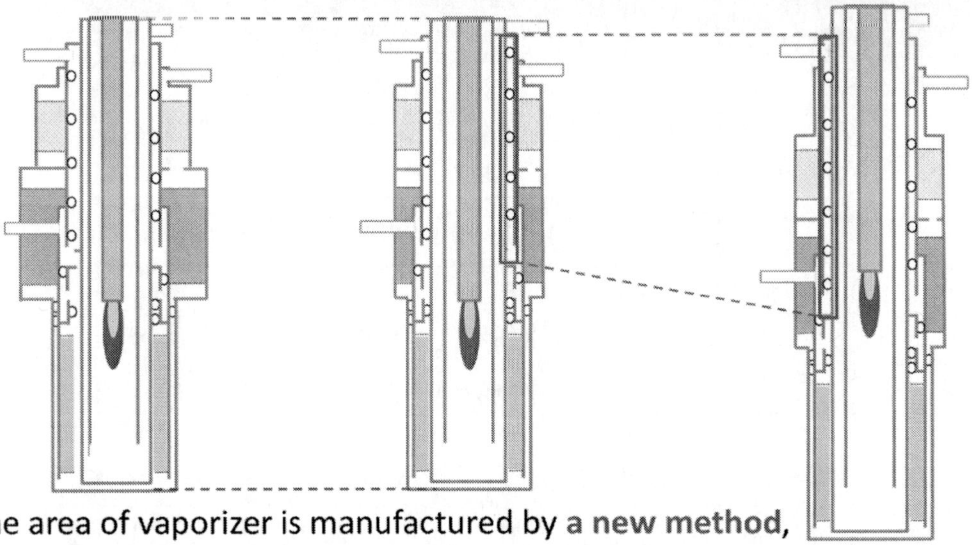

If the area of vaporizer is manufactured by **a new method**,

⇒ The flame is positioned at the same height as the present FP.

⇒ Vaporization capability is the same as the high-S/C FP.

TOKYO GAS 15

New manufacturing method of vaporizer

Previously method

- Forming spiral flow path using round metal bars.
- After spot-welding the round metal bar to the inner tube, the outer tube is fitted.

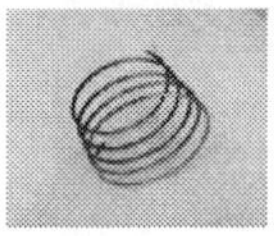

New method

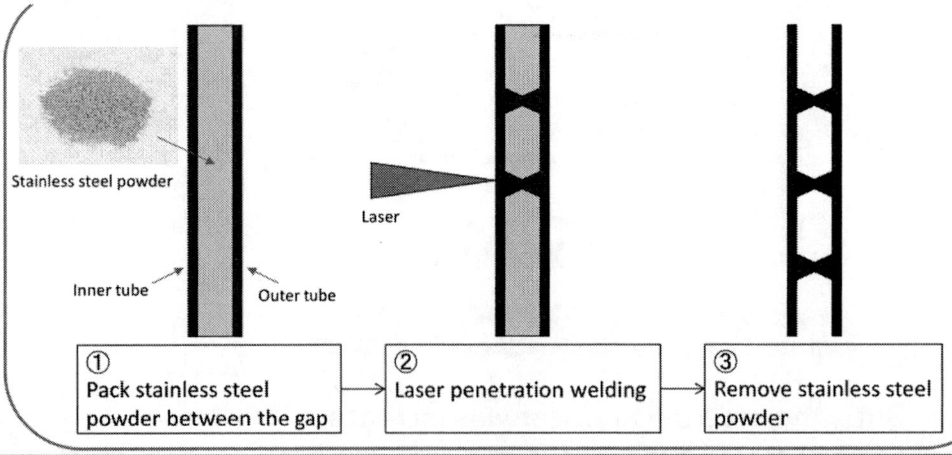

TOKYO GAS 16

New manufacturing method of vaporizer

	Previous method	New method
materials	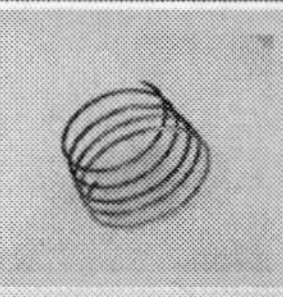	
cross-section view	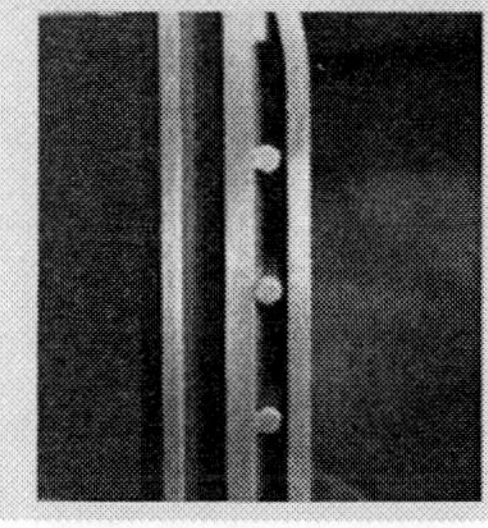	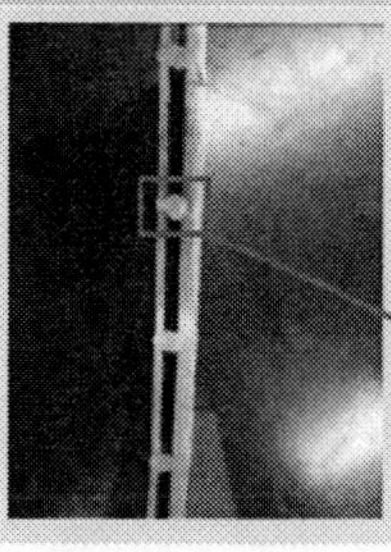

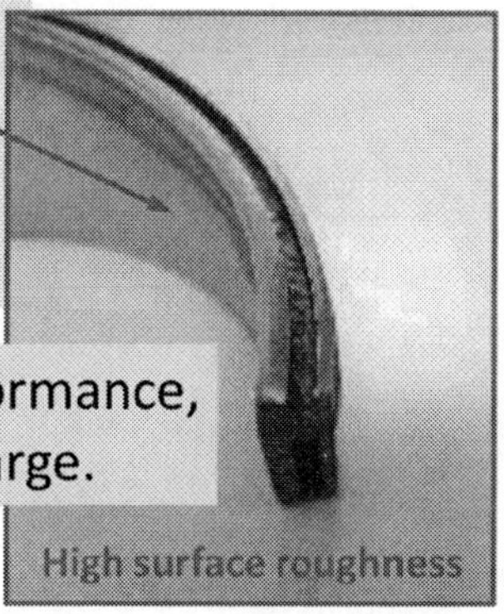

The vaporizer can expect high heat exchange performance, since the surface area of the flow path surface is large.

FP performance testing equipment

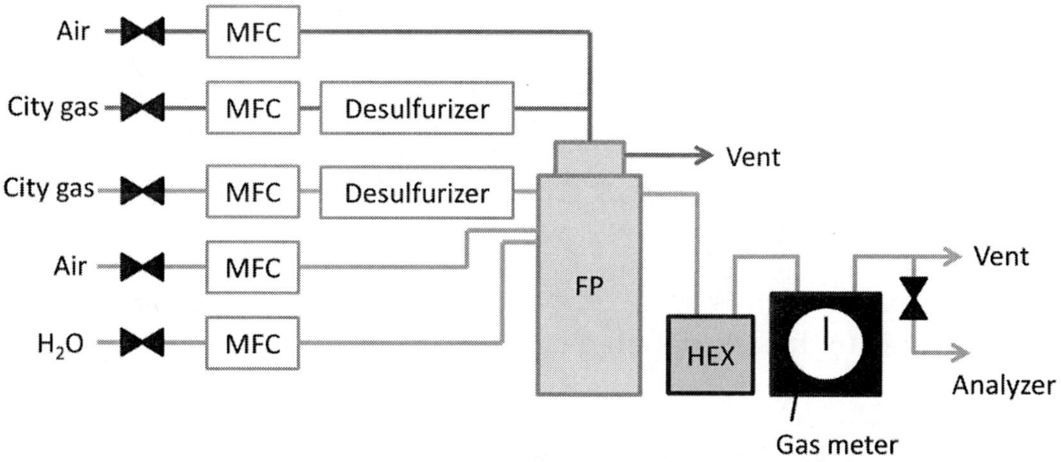

Analyzer : GC-TCD, IRA

Performances of the high-S/C FP

		Calculation #1	High-S/C FP	Advanced High-S/C FP
Condition	S/C	4	3.9	3.9
	Outlet temp. of SR catalyst	580°C	570°C	570°C
Performance	CH_4 conversion	76%	74%	74%
	CO concentration at SR catalyst	5.30%	-	-
	H_2 generation efficiency	80%	80%	79%
	H_2 utilization ratio	85%	86%	85%
	CO concentration at PROX catalyst	-	<10ppm	<10ppm

- CO concentration at PROX catalyst was less than 10ppm.
- H_2 generation efficiency of the high-S/C FP and the advanced high-S/C FP corresponded to the calculation result.

✓ Operating condition to reduce the amount of WGS catalyst was established.
✓ High-S/C operation is possible with the same size as the present FPS.

TOKYO GAS 19

Summary

☐ To reduce the FP cost, we focused on reducing the amount of WGS catalyst.

☐ The amount of WGS catalyst was decreased by high S/C and low outlet temperature of SR catalyst.

☐ We designed an advanced high-S/C FP for high-S/C operation with the same size as present FP.

☐ The performance of the high-S/C FP and the advanced high-S/C FP corresponded to the calculation results.

☐ The FP cost could be reduced by 20% or more.

A policy on design of FP
for further cost reduction was established.

TOKYO GAS 20

Thank you for your attention.

17 AIChE Spring Meeting, San Antonio, TX

13TH GLOBAL CONGRESS ON PROCESS SAFETY

TOKYO GAS

LAMAR UNIVERSITY
MEMBER THE TEXAS STATE UNIVERSITY SYSTEM

Synergy of Large-Scale Cryogenic Air Separation Unit and Liquefied Natural Gas Import Terminal

Presented by
Helen Lou, Ravinder Singh.*

Dan F. Smith Department of Chemical Engineering,
Lamar University, Beaumont, Texas.

AIChE Spring 2017 Meeting

Energy generation outlook (2005 - 40)

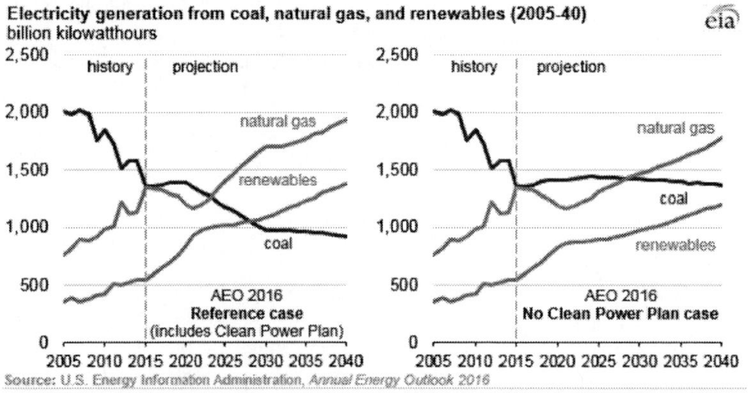

Figure 2: Electricity generation trends

Consumption of Natural Gas

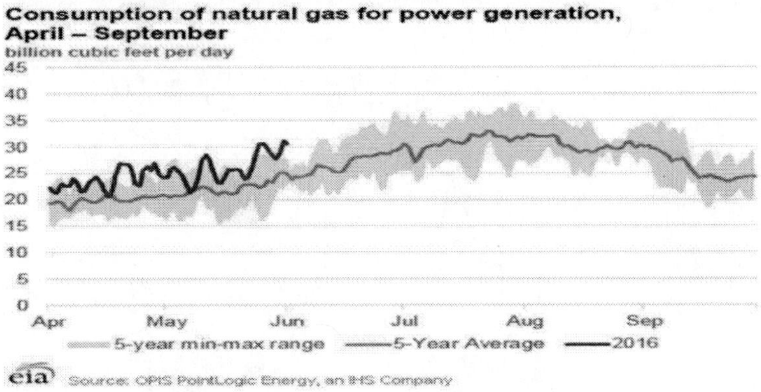

Figure 3: Trends of natural gas consumption

LNG Import Terminal

Figure 4: LNG Terminal (Image source: Tokyo Gas)

- LNG imported using insulated tankers.
- The tankers unload the LNG into large insulated storage tanks
- LNG is then pumped to required pressure and regasified for final transmission through the pipeline to the consumer

Basic LNG Terminal

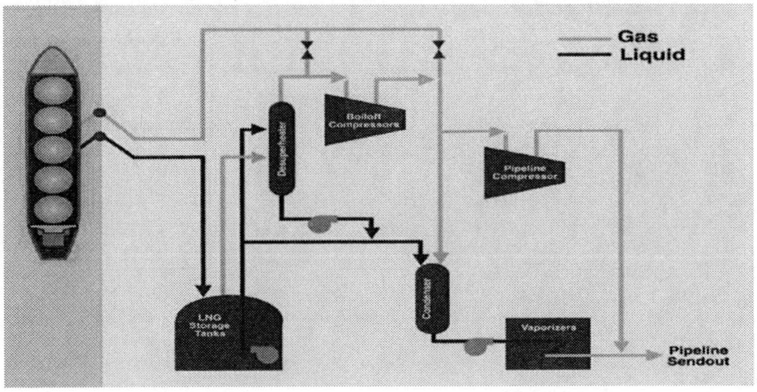

Figure 5: A basic LNG terminal (Hydrocarbons Technology Inc.)

Common LNG vaporization technologies

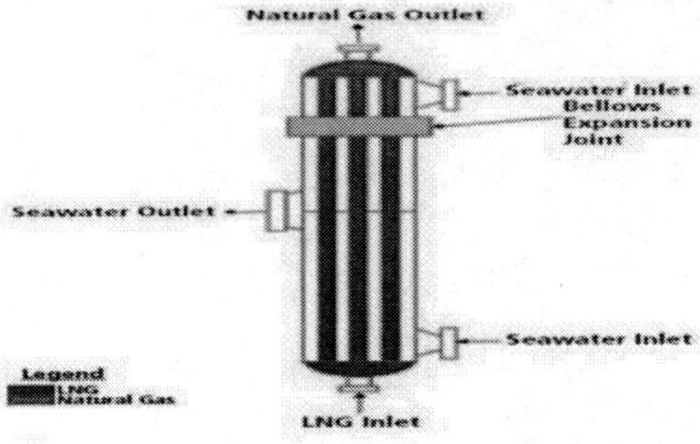

Figure 8: Shell and tube vaporizer (source: www.uscg.mil)

Common LNG vaporization technologies

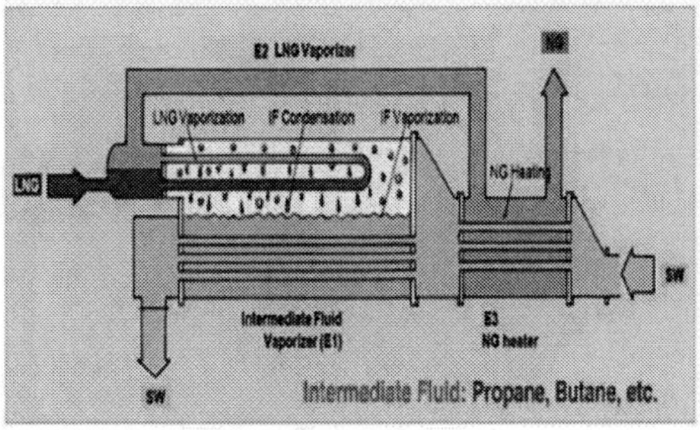

Intermediate fluid vaporisers (IFV)

Figure 9: Intermediate fluid vaporizer (source: www.uscg.mil)

LNG Import terminal HYSYS simulation

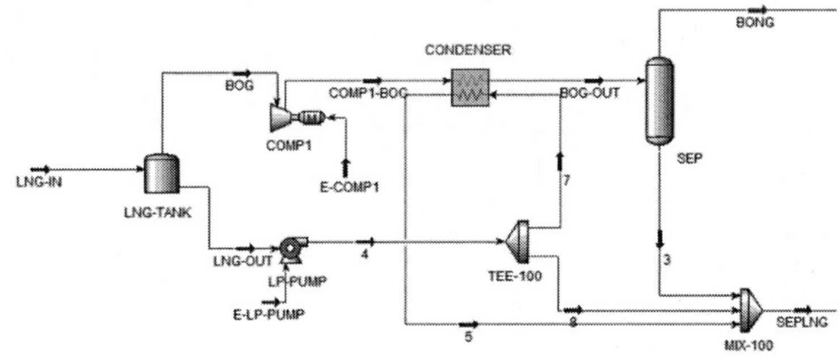

Figure 10: LNG Import Terminal

LNG Import terminal HYSYS simulation

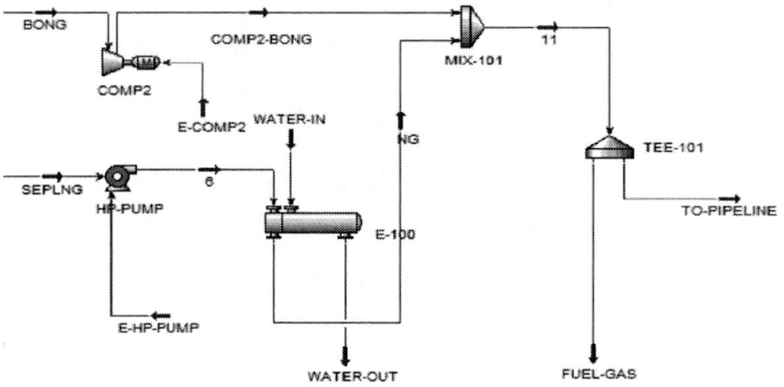

Figure 11: LNG Regasificaiton process by water

LNG Terminal simulation stream results

Table 1: Selected material streams

Material Streams | Compositions | Energy Streams | Unit Ops

Name	LNG-IN	BOG	LNG-OUT	BOG-OUT	COMP2-BONG	COMP1-BOG	BONG
Vapour Fraction	0.0300	1.0000	0.0000	0.0500	1.0000	1.0000	1.0000
Temperature [K]	114.5	114.5	114.5	158.2	247.1	250.3	158.2
Pressure [bar]	1.216	1.216	1.216	15.00	60.00	15.20	15.00
Molar Flow [MMSCFD]	743.6	22.31	721.3	22.31	1.115	22.31	1.115
Mass Flow [tonne/d]	1.500e+004	427.8	1.457e+004	427.8	21.39	427.8	21.39
Liquid Volume Flow [m3/h]	2046	59.54	1987	59.54	2.977	59.54	2.977
Heat Flow [MW]	-929.4	-25.06	-904.3	-26.63	-1.208	-23.75	-1.242

Name	WATER-IN	WATER-OUT
Vapour Fraction	0.0000	0.0000
Temperature [K]	308.1	278.1
Pressure [bar]	1.216	1.016
Molar Flow [MMSCFD]	4213	4213
Mass Flow [tonne/d]	9.072e+004	9.072e+004
Liquid Volume Flow [m3/h]	3788	3788
Heat Flow [MW]	-1.837e+004	-1.870e+004

Table 2: Energy stream results

Material Streams | Compositions | Energy Streams | Unit Ops

Name	E-COMP1	E-LP-PUMP	E-COMP2	E-HP-PUMP
Heat Flow [MW]	1.312	0.7199	3.390e-002	2.436

Cryogenic Distillation of air

- Recovering the refrigeration of product gases :
  - Recuperative heat exchange is used to recover the refrigeration in un-liquefied product gases.
- Using gas compression and subsequent expansion :
  - The enthalpy of gas, in this case air, is decreased by isothermal compression. This can be achieved by compression followed by aftercooling.
  - The enthalpy of air is further decreased by an expander, once it has been isothermally compressed.

Cryogenic Distillation of Air

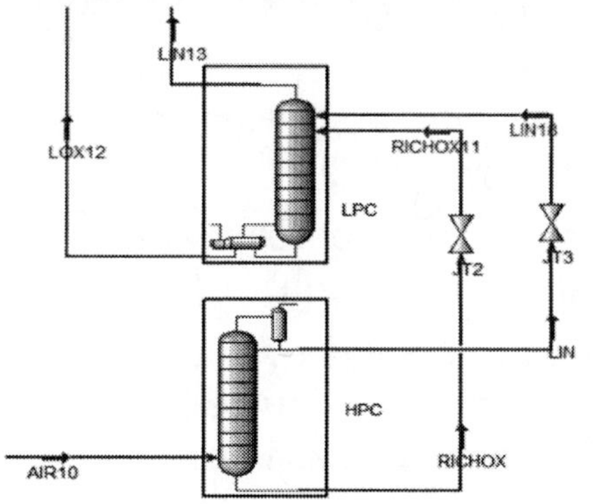

Figure 12: Linde double column system

- Double column system having a common reboiler and condenser is the heart of the air separation process.
- High Pressure Column (HPC) operates between 5 bar and 7 bar.
- Low Pressure Column (LPC) operates between 1.3 bar and 1.8 bar.

Comparison of air liquefaction system configurations

Table 3:Comparison of air liquefaction systems (Source: Cryogenic Systems, 2nd Edition, Randall F. Barron

	Air liquefaction system	Liquid yield, $y = m_f/m$	Work per unit mass liquefied, W/m_f (BTU/lb$_m$)
1	Linde Dual Pressure System (p_2 = 60 atm, p_3 = 200 atm)	0.057	1314
2	Claude System ($x = m_e/m = 0.70$)	0.262	350
3	Heylandt System (p_2 = 200 atm, x = 0.60)	0.377	**275**

- T_1 = 300 K and p_1 = 1 atm.; Expander efficiency = 80 %

Heylandt System (Overview)

- Double column system is used for cryogenic distillation.
- Products from this system are usually liquid nitrogen (LIN) and liquid oxygen (LOX).
- Incoming air is compressed to approximately 200 bars through multi-stage compression.
- For this system, the optimum value of temperature before expander is near ambient, thus eliminating the need for third heat exchanger, as in the case of Claude system.

Heylandt System HYSYS Simulation

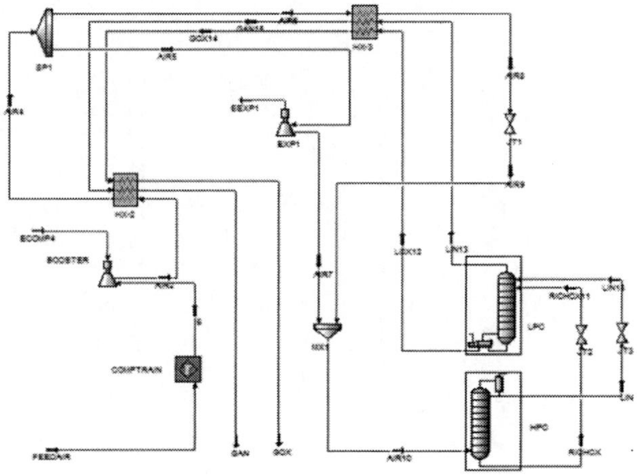

Figure 13: Heylandt System for air separation

Heylandt System HYSYS Simulation

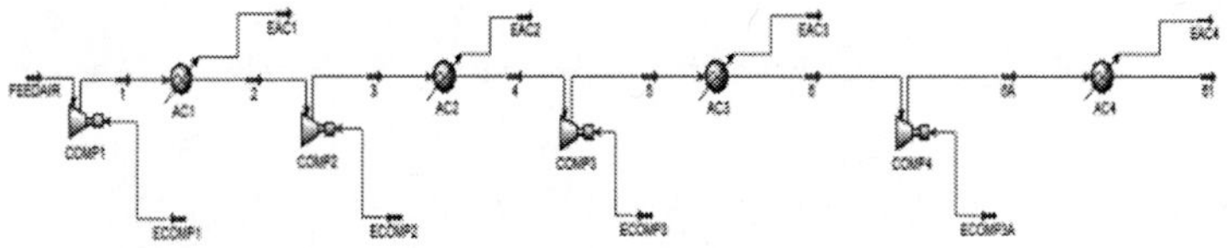

Figure 14: Compressor train

Table 4 : Material streams in the compressor train

	Unit	FEEDAIR	1	2	3	4	5	6	6A	61
Vapour Fraction		1	1	1	1	1	1	1	1	1
Temperature	K	315.00	475.10	310.00	432.30	310.00	513.69	310.00	401.89	315.00
Pressure	bar	1.01	3.20	3.10	7.78	7.68	32.07	31.97	65.00	64.90
Molar Flow	MMSCFD	433.29	433.29	433.29	433.29	433.29	433.29	433.29	433.29	433.29
Mass Flow	tonne/d	15000	15000	15000	15000	15000	15000	15000	15000	15000
Liquid Volume Flow	m3/h	718.66	718.66	718.66	718.66	718.66	718.66	718.66	718.66	718.66
Heat Flow	MW	2.90	31.37	1.93	23.56	1.73	38.10	0.70	16.87	0.40

Heylandt system with LNG cold stream

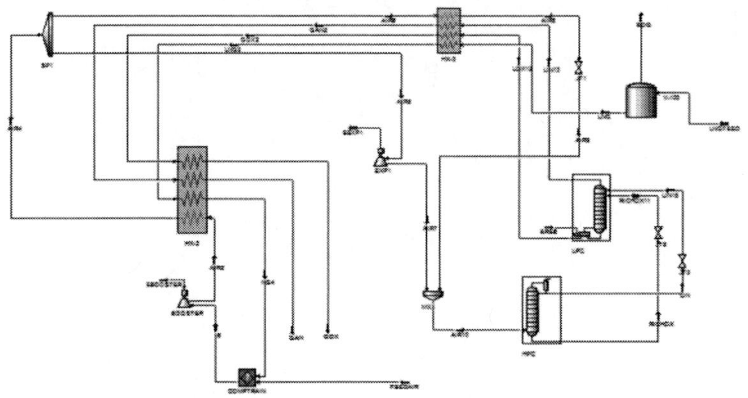

Figure 19: Heylandt System with LNG cold stream

Heylandt system with LNG cold stream

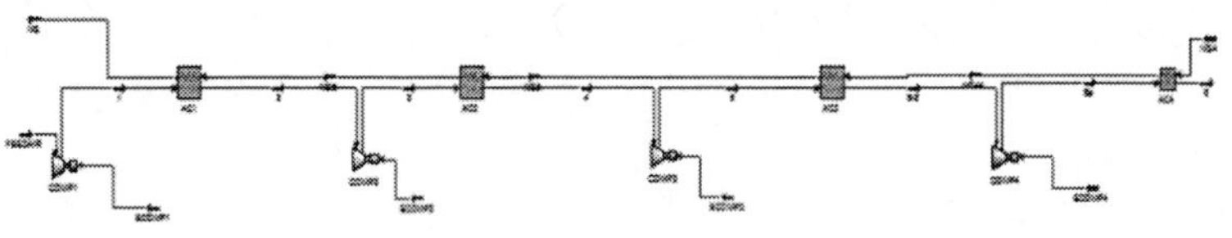

Figure 20: Compressor train

Table: Selected material streams

Material Streams | Compositions | Energy Streams | Unit Ops

Name	FEEDAIR	NG4	6	NG	5	NG5	1
Vapour Fraction	1.0000	0.4583	1.0000	1.0000	1.0000	0.9941	1.0000
Temperature [K]	315.0	117.9	310.0	296.2	636.1	155.1	455.6
Pressure [bar]	1.013	1.500	64.90	1.100	64.14	1.300	2.838
Molar Flow [MMSCFD]	433.3	743.6	433.3	743.6	433.3	743.6	433.3
Mass Flow [tonne/d]	1.500e+004	1.500e+004	1.500e+004	1.500e+004	1.500e+004	1.500e+004	1.500e+004
Liquid Volume Flow [m3/h]	718.7	2046	718.7	2046	718.7	2046	718.7
Heat Flow [MW]	2.897	-892.0	-0.5663	-778.2	60.91	-830.2	27.85

Table: Energy Streams of COMPTRAIN Section

Material Streams | Compositions | Energy Streams | Unit Ops

Name	ECOMP1	ECOMP2	ECOMP3	ECOMP4
Heat Flow [MW]	24.95	25.89	59.20	0.3023

Results of LNG Terminal

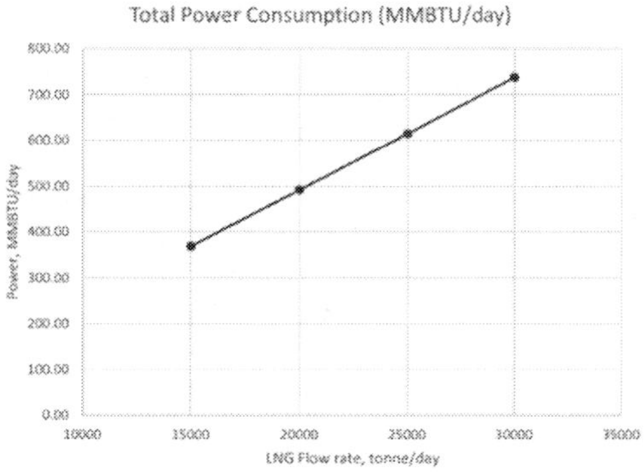

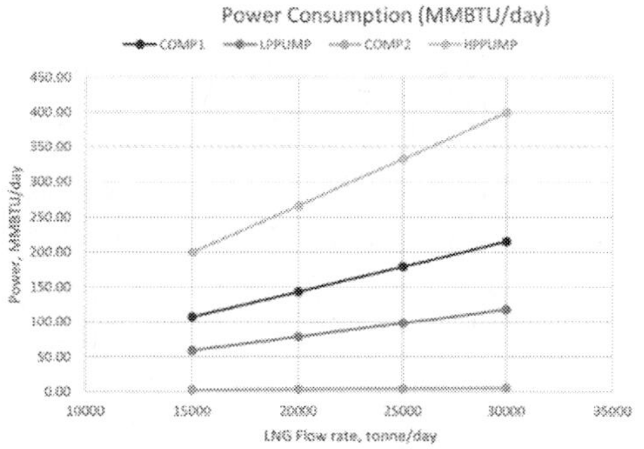

Air Separation Unit results

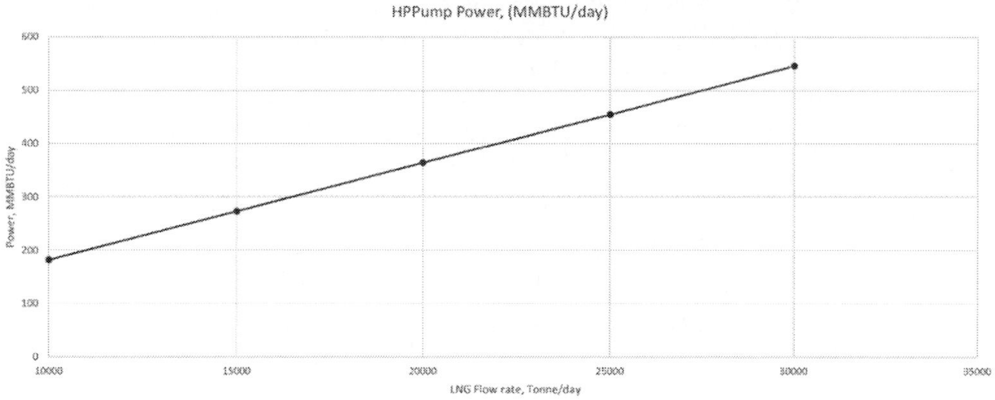

Conclusion

- For 15000 TPD of LNG receiving terminal, power required to regasify LNG is 368.576 MMBTU/day
- For Air Separation Unit utilizing 15000 TPD of LNG, power required to regasify LNG is 272.8 MMBTU/day
- There is 95.776 MMBTU/day of power saving to regasify LNG.

References

- Barron, R. (1985). *Cryogenic Systems* (Second Edition ed.). Oxford University Press, New York.
- Dawson, B., Siegmund, S., & Yonggui, Z. (2004). Flowsheet Optimization for multi-product air separation units. *The First Baosteel Annual Academic Conference, Shanghai, China.*
- Kawamoto, H. (2008). *Natural gas regasification technologies: Efficient environmentally friendly LNG vaporization methods.*
- Latimer, R. (1967). Distillation of Air. *Chemical Engineering Progress*, 35-59.
- Smith, A., & Klosek, J. (2001). A review of air separation technologies and their integration with energy conversion process. *Fuel Processing Technology*, 115-134.

Questions ?

Pros and Cons of Different Blanket Gas Control for Different Unit for LNG Baseload Plant.

Dr. Yoshitugi Kikkawa,

Doctor in Engineering,

Fellow AIche

Kikkawa Giken, Japan

J C Kuo, Ph.D. , P.E.

University of Houston

Aiche, Spring Meeting ,

March 29, 2017, San Antonio, Texas

Pros and Cons of Different Blanket Gas Control for Different Unit for LNG Baseload Plant.

Dr. Kikkawa is working as the Representative of Kikkawa Giken. He has been working as a senior engineering consultant for many mega LNG Projects in Chiyoda Corporation. He has a Bachelor degree of Fuel Chemistry from Akita University, He has a Doctor degree of Engineering from Kyushu University.

Major project experiences; Japanese's Grass Roots Refinery Projects, Mega LNG Projects.

Member: Fellow, American Institute of Chemical Engineers

J.C. is a veteran of the Gas Processing/LNG industry, with 40 years experience. He has served as the Process Manager for many projects. He has a BS from Chung Yuan Christian University, Taiwan and a MS from the University of Houston, both for Chemical Engineering. He also holds a MS in Environmental Engineering from Southern Illinois University. He has PH. D from University of Houston and a registered Professional Engineer in Texas.

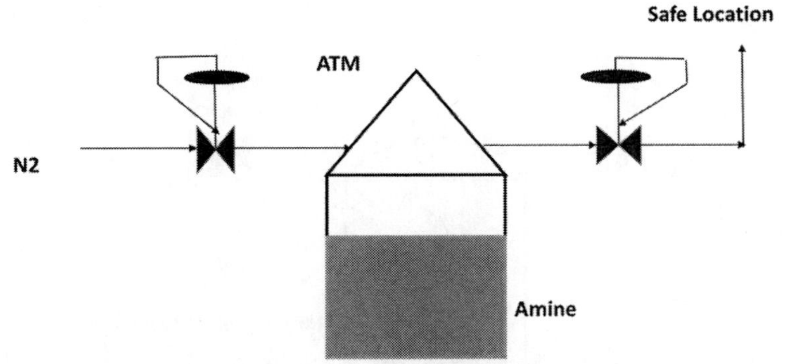

1. **Amine Process Surge tank (40% to 50% max. weight solution)**
2. **Amine field storage tank (nearly 99% weight solution)**
3. **Self regulating control valves or DCS split range control valves.**
4. **Prevent Amine degradation (O2 in PPM level) to form HSS. N2 used to control to near ATM pressure. Vent to safe location.**

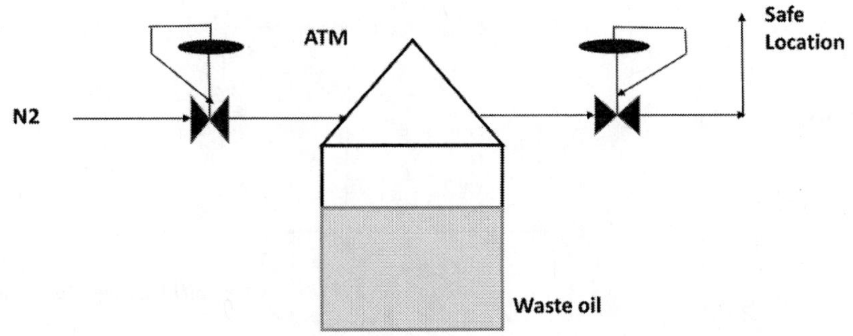

1. **Waste oil storage tank (From CPI , Oil and Water Separator)**
2. **Self regulating control valves or DCS split range control valves.**
3. **Explosion Prevention. N2 used to control to near ATM pressure. Vent to safe location.**

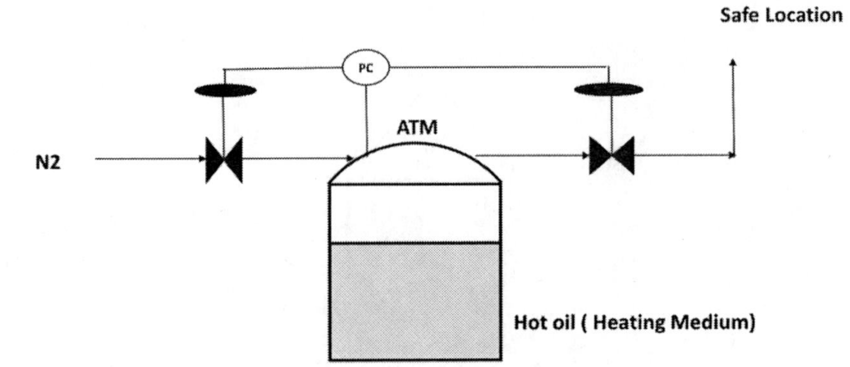

1. Hot oil field storage tank (nearly 99% weight solution)
2. Self regulating control valves or DCS split range control valves.
3. Explosion Prevention. Prevent Heating medium (hot oil) degradation.
4. N2 used to control to near ATM pressure. Vent to safe location.

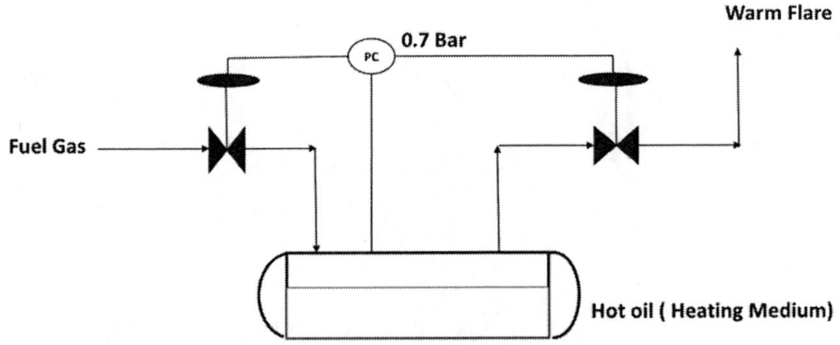

1. Hot oil Process surge tank (nearly 99% weight solution)
2. DCS split range control valves at slightly high pressure of 0.7 Bar.
3. Prevent NPSH problem for oil circulating pumps.
4. Fuel gas used to control to near 0.7 Bar. Vent to Warm flare.

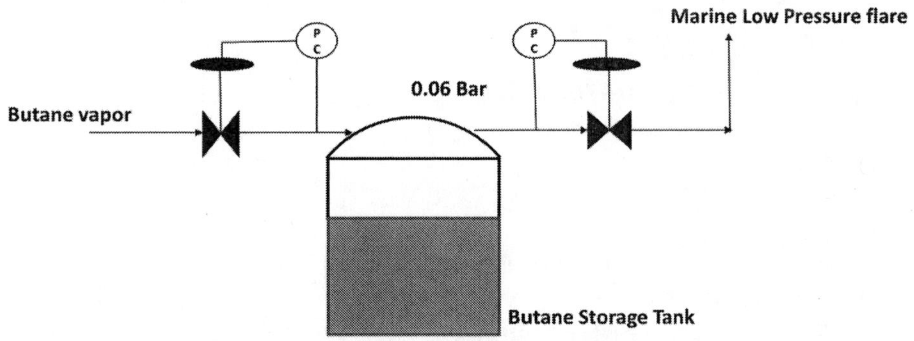

1. **Butane storage tank (nearly -4.4 C and 0.06 Bar)**
2. **DCS with two PC's and two control valves, one from tanker.**
3. **Tank protection due to pressure. Butane vapor used to control to near ATM pressure. Vent to marine low pressure flare.**

1. **LNG storage tank (nearly -160.2 C and 0.05 Bar)**
2. **DCS with two PC's and two control valves.**
3. **BOG compressors provided with FC and PC to control tank pressure.**
4. **Tank protection due to pressure. Fuel gas vapor (HP or LP) also used to control to near ATM pressure. Vent to marine low pressure flare.**
5. **Propane Storage tank has similar set up with BOG Comp.'s**

Self regulating control valves or DCS split range control valves ?

(1) Self regulating control valves

1. More economical and lower CAPEX
2. Workable for
   - low cost, carbon steel,
   - smaller and
   - low design pressure tank
   - Low frequency used tanks such as offsite amine or hot oil storage tank.

3. Need to pass through HAZOP

(2) DCS split range control valves ?

1. Easier for operation with control room operator's accesses and alarm.
2. Higher cost.
3. For big and expensive thank such as LNG , Butane and Propane storage tank.
4. For critical utilities such Hot Oil system.

Pressure control set points

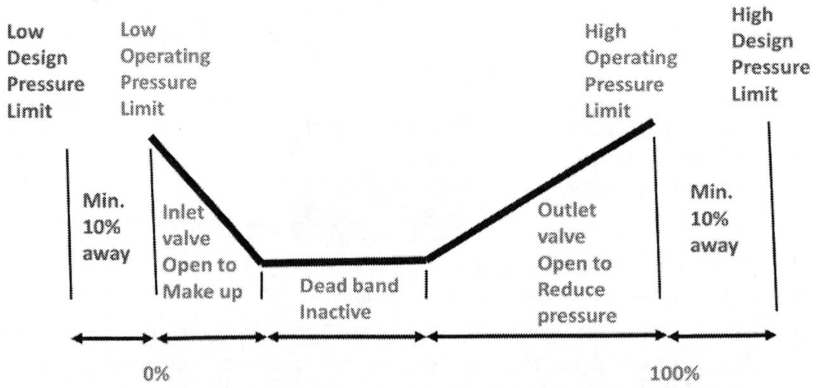

Other actual blanketing gas system

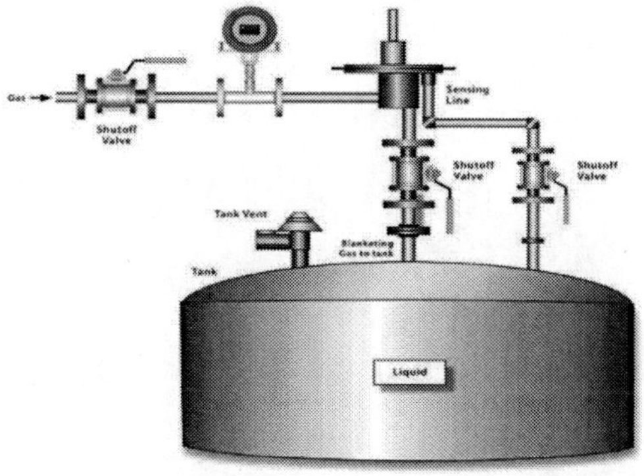

Sensing line work as pressure control to introduce gas to do blanketing.

Pros and Cons of Different Blanket Gas Control for Different Unit for LNG Baseload Plant.

Thank you very much !

Q and A , please

ID# 480251

THE LINDE GROUP

Linde

How Non-Condensables Affect Your Ethane/Ethylene Liquefaction Process

Concepts to Reduce Power Consumption and Product Losses

Thomas Walter
Kathy Bigger
Dr. Adrian Wegmann

Linde Engineering North America Inc.

Why do we worry about non-condensable gases?

THE LINDE GROUP

Linde

- Inert gases which are not soluble in the product may significantly affect the performance of a closed loop refrigeration system
- Unlike for LNG plants, this issue may become far more severe for C2 Liquefaction processes
  - ➢The solubility for certain inerts in liquid Methane is different for liquid Ethane/Ethylene
  - ➢Typically, flash and boil-off gas cannot be compressed and sent back to pipeline/fuel gas
- Typical critical components in the feed are: H_2, N_2, O_2, CH_4

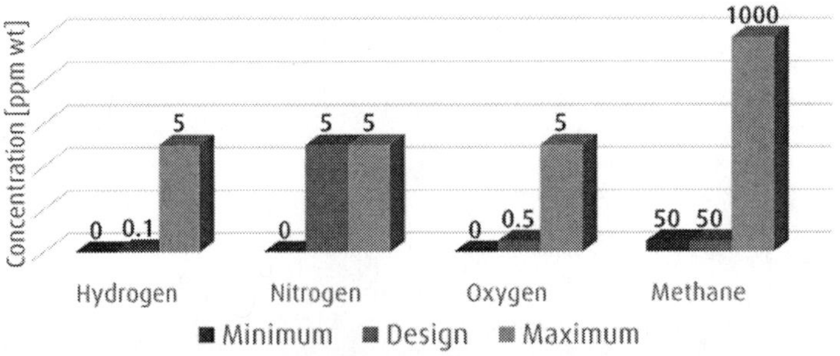

Chart 1: Design case study, minimum and maximum concentration in polymer grade Ethylene (>99.9%)

Which gas is most critical?

- Comparison of K-values for different lighter contaminants in liquid Methane, Ethane and Ethylene:

	Methane	Ethylene	Ethane
Hydrogen	1600	1350	2070
Nitrogen	24	235	260
Methane	-	24	32
Oxygen	5	58	82

Table1: Comparison of K-values (PengRobinson) at atmospheric storage conditions

→ Hydrogen together with Nitrogen are most critical

Mixed refrigerant liquefaction process
LIMUM®1 (LInde MUltistage Mixed Refrigerant)

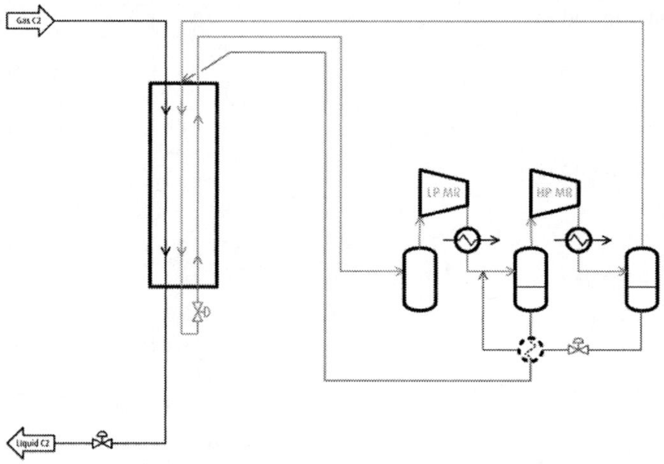

Single Cycle Process

- Typical for peak shaving or small scale LNG plants; process adapted to liquefaction of C2
- Simple process, low equipment count
- One refrigerant cycle providing refrigeration duty
- Pre-cooling, liquefaction and sub-cooling cycle realized with a Plate-Fin Heat Exchanger (PFHE)
- Lower specific energy consumption compared to an expander process

Recompression to feed pressure
Concept 1

- Flash and tank boil-off gas is compressed to feed pressure
- Simple design
- Build-up of light components due to recycle
- Purge rate can be adjusted depending on inlet concentration
- Purge to be sent to vent, flare or fuel
  - ➢Environmental restrictions possible

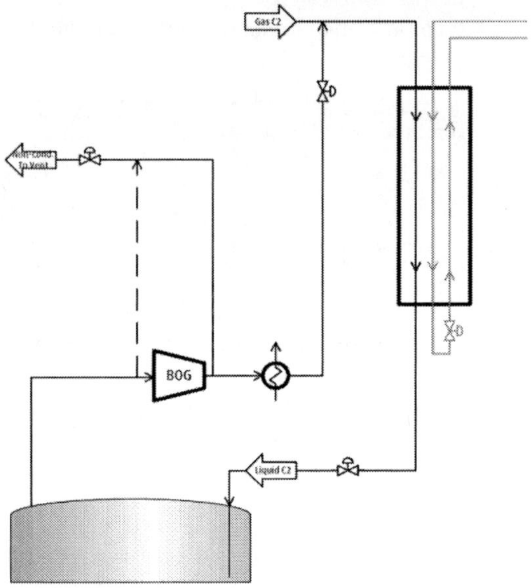

ID# 480251 : How Non-Condensables Affect Your Ethane/Ethylene Liquefaction Process - Different Concepts to Reduce Power Consumption and Product Losses
Linde Engineering North America, Inc. - Kathy Bigger, Thomas Walter, Adrian Wegmann

5

Recompression to feed pressure with vapor recovery unit
Concept 2

- Flash and tank boil-off gas is compressed to feed pressure
- Simple design
- Build-up of light components due to recycle
- Purge rate can be adjusted depending on inlet concentration
- Purge to be sent to vent, flare or fuel
  - ➢Environmental restrictions possible
- Can be equipped with VRU to recover product losses

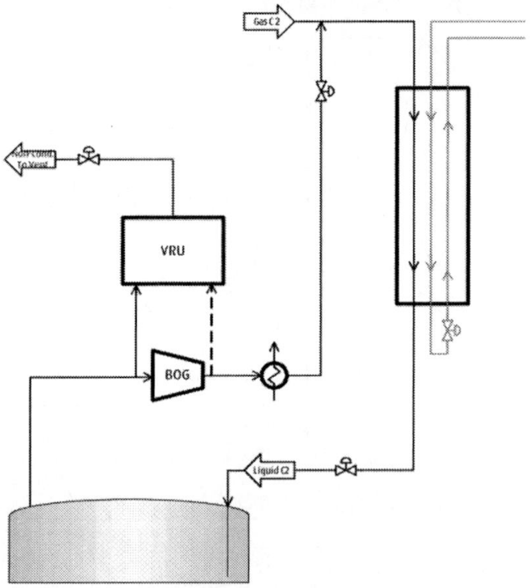

ID# 480251 : How Non-Condensables Affect Your Ethane/Ethylene Liquefaction Process - Different Concepts to Reduce Power Consumption and Product Losses
Linde Engineering North America, Inc. - Kathy Bigger, Thomas Walter, Adrian Wegmann

6

Recompression to intermediate pressure
Concept 3

- Flash and tank boil-off gas is being compressed to intermediate pressure
- Re-liquefaction below critical pressure in separate PFHE pass
  - ➢Reduced flash and boil-off gas compression
- Purge taken from intermediate flash drum
  - ➢Additional vessel required
  - ➢Reduced product losses

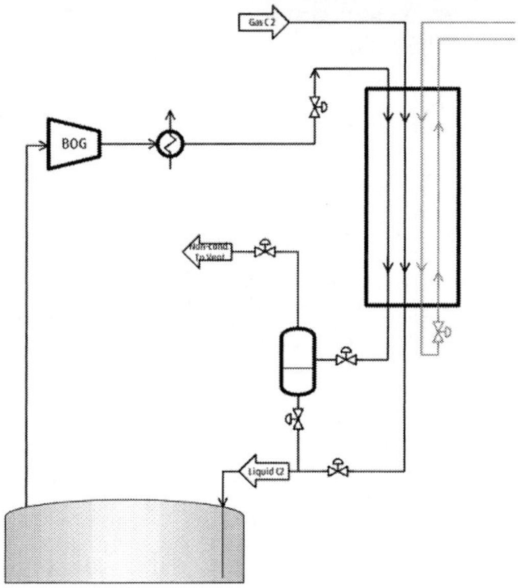

Case study

Feed:

- Polymer grade Ethylene (>99.9%)
- 65 bar pipeline pressure
- Liquefaction with storage at atmospheric pressure

	Design Case	Rating Case
Hydrogen	0.1 ppm wt	5 ppm wt
Nitrogen	5 ppm wt	5 ppm wt
Methane	50 ppm wt	300 ppm wt
Oxygen	0.5 ppm wt	1.0 ppm wt

Table2: Concentration of inerts for design and rating case

Overview results

Design Case [1]	Concept 1	Concept 2	Concept 3
Power consumption [2]	±0%	±0%	+1-2%
Vent rate	2.1 kg/h	<0.1 kg/h	<0.1 kg/h
Product loss	2.0 kg/h	<0.1 kg/h	<0.1 kg/h
Nitrogen consumption [3]	-	7.4 kg/h	-

Rating Case [1]	Concept 1	Concept 2	Concept 3
Power consumption [2]	±0%	±0%	+1-2%
Vent rate	132 kg/h	<1.0 kg/h	<3 kg/h
Product loss	130 kg/h	-	<1.5 kg/h
Nitrogen consumption [3]	-	485 kg/h	-

General: Numbers shown are calculated only. No margin etc. added or heat losses considered.
1) Normalized to a feed rate of 100 t/h 2) Simulated power consumption using 80% polytropic efficiency 3) Liquid Nitrogen stored at 5 bar, utilizing latent heat only

Composite heat curve PFHE

Concept 1 & 2

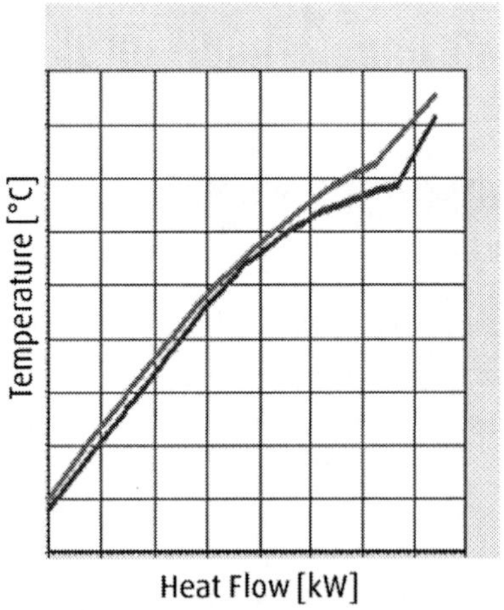

Concept 3

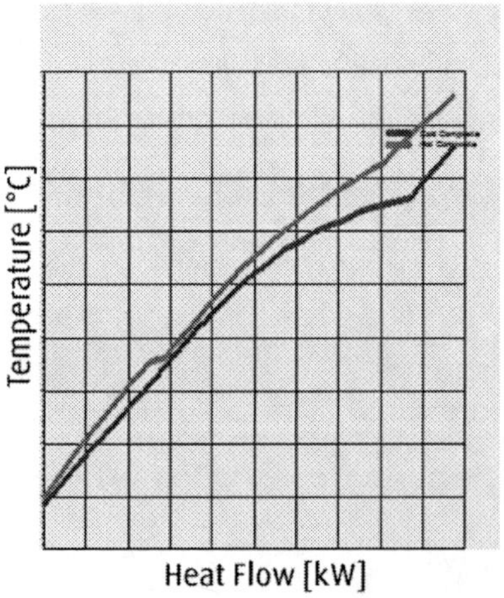

Remark: Effect of discontinuity oversaturated for illustration purpose

Cost comparison

Design Case 1)	Concept 1	Concept 2	Concept 3
Electricity 2)	±0	±0	+$53,000
Product loss 3)	$12,000	-	-
Nitrogen costs 4)	-	$3,800	-
Total	**+$12,000**	**+$3,800**	**+$53,000**

Rating Case 1)	Concept 1	Concept 2	Concept 3
Electricity 2)	±0	±0	+$56,000
Product loss 3)	$782,500	-	$9,200
Nitrogen costs 4)	-	$247,600	-
Total	**+$782,500**	**+$247,600**	**+$65,200**

1) Normalized to a feed rate of 100 t/h and 8500h/year
2) Electricity costs of 0.07 $/kWh
3) Mont Belvieu Ethylene price January 2017 (708.60$/t)
4) Nitrogen price of 60 $/t

Concept 2: Cryogenic Condensation
CIRRUS®

- Standardized cryogenic condensation units to recover VOC's using LIN
- Available in different sizes
- Flexible solution depending on inlet inert concentration

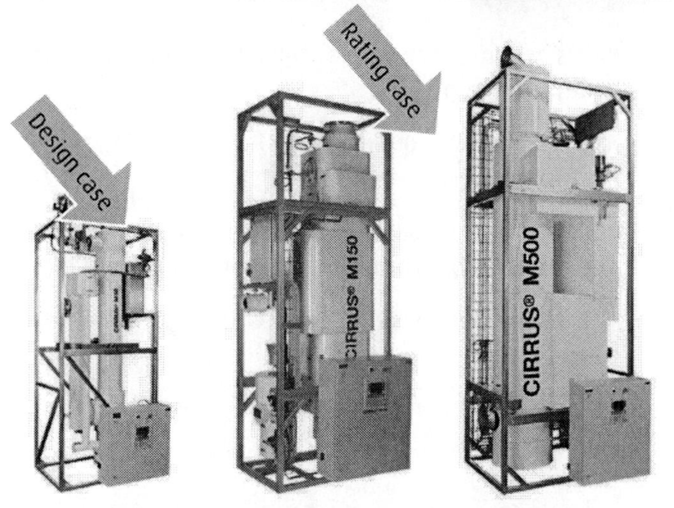

CIRRUS® M50, M150 and M500 modules. The CIRRUS® M50 module is equipped with an optional valve.

http://www.lindeus.com/en/processes/freezing_and_cooling/cryogenic_condensation/index.html

Concept overview

THE LINDE GROUP

Linde

Concept 1: Compression to feed pressure – Purge to Vent, Flare or Fuel

+	–
• Simple concept • Normal operating costs at low inert concentrations	• Environmental restrictions possible • High product loss at high inert concentrations

Concept 2: Compression to feed pressure – Purge to VRU

+	–
• Flexible concept with minimized product loss • Normal operating costs at low inert concentrations	• Higher operating costs with higher inert concentrations • Additional equipment (LIN storage, VRU etc.) required

Concept 3: Recompression to intermediate pressure

+	–
• Minimized product loss • Operating costs independent from inert concentration	• Slightly Higher operating costs at low inert concentration • More complex process/PFHE design

Summary

THE LINDE GROUP

Linde

- Inerts or non-condensables can have a significant impact on Ethane/Ethylene liquefaction processes
- Hydrogen and Nitrogen are most critical and their concentration significantly affects the process design
- Purging to vent, flare or fuel may not be permitted due to environmental constraints and results in high product losses at higher inert inlet concentration
- A cryogenic condensation unit/VRU can be used flexible at variable inert inlet concentrations
- The re-liquefaction at intermediate pressure is most robust and suitable for all inlet concentrations

→ Optimal concept to be identified and designed in collaboration with client

Thanks for your attention.

Vielen dank für Ihre Aufmerksamkeit.

Thomas.Walter@Linde.com

ENHANCED SAFETY IN LNG PLANT DESIGN

Augusto Bulte
LNG/Gas Monetization Manager
&
Lourdes Martinez
Principal Process Engineer

Amec Foster Wheeler
585 N Dairy Ashford, Houston – Texas 77079

AIChE Spring Meeting and 13th Global Congress on Process Safety, San Antonio - Texas

Abstract

LNG plants are designed based on N+1, or even N+2, philosophy to provide high availability. Based on the same philosophy, why we don't apply this criterion in safety?

Amec Foster Wheeler is including in their design some depressurization systems in places where there is high pressure or high volume of Liquefied Natural Gas (LNG) (or Natural Gas (NG)) to provide more safety, despite the fact that API does not consider it mandatory. Such design provides more safety, allows recovery of LNG and reduces the flare emissions in case of overpressure.

Introduction

LNG plants are usually designed following specific standards, namely: NFPA 59A, API 521, EN 1473, CSA Z276,..., and OSHA (PSM). These specifications provide us the minimum requirements in terms of safety. Plants are designed following these minimums, although usually there is no smart design in terms of process safety.

Value Engineering is usually performed after finalizing a FEED. Following that principle some sort of "Safety Value Engineering" should follow a HAZOP session.

No matter of line of work you are in, the improvement in safety should apply and not just to stick to minimum recommendations. Codes, norms, specifications and standards always define the safety requirements that are to be applied as "at least". Obviously, in most of cases, the designs of plants follow these minimum requirements, but: are plants truly designed following the safety criteria? or are they just a compliance of the minimum requirements as described in the applicable standards?

LNG plants

The LNG supply chain from gas fields to consumers comprises a number of processes and logistical steps. Natural gas is obtained from field, impurities are removed and it is liquefied to be transported due to the fact that the liquefied natural gas needs 600 less space than the natural gas in his own stage. The LNG is transported by LNG carriers to the LNG import and regasification terminals.

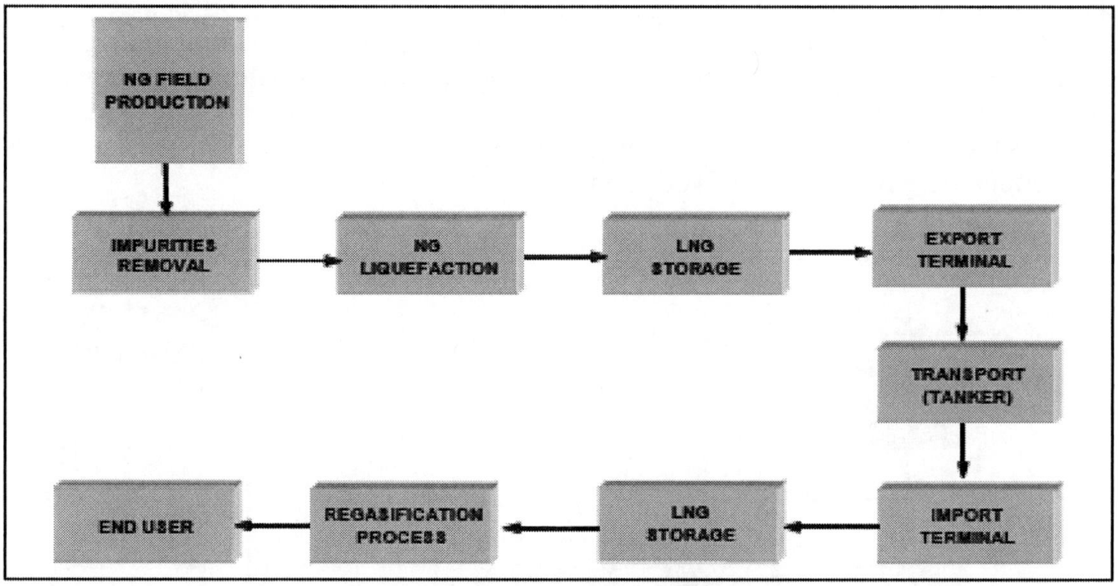

Figure 1. LNG supply chain

Regasification Terminal

Terminals are intended for the reception and storage of LNG from carriers. This LNG will be pumped-out from storage tanks, vaporized into NG and sent to the transmission system at required conditions of pressure and temperature. The following Block Process Flow Diagram can be used to summarize the process. Nowadays, there are several LNG terminal types: offshore, onshore, near offshore… in which the main characteristic is their location.

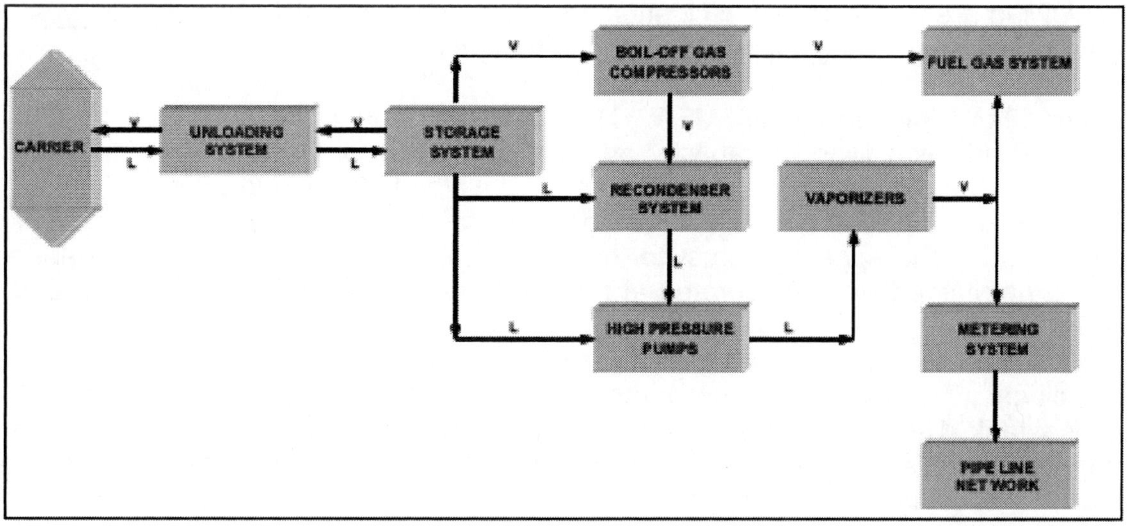

Figure 2. Block Process Flow Diagram. Regas process

Liquefaction Plant

The liquefaction process takes place in a liquefaction plant. In this process the natural gas is fed into the gas liquefaction plant (LNG Plant) being previously treated. The gas treatment involves removal of certain components, such as dust, acid gases (hydrogen sulfide, carbon dioxide), helium, water, and heavy hydrocarbons (e.g., benzene), and other components that are subjected to freeze, which could cause difficulty downstream or be destructive to the liquefaction facility. The sweetened gas is later on liquefied. All the liquefaction processes are based on the reverse Rankine cycle, where refrigeration duty is supplied by evaporation of liquid phase refrigerants.

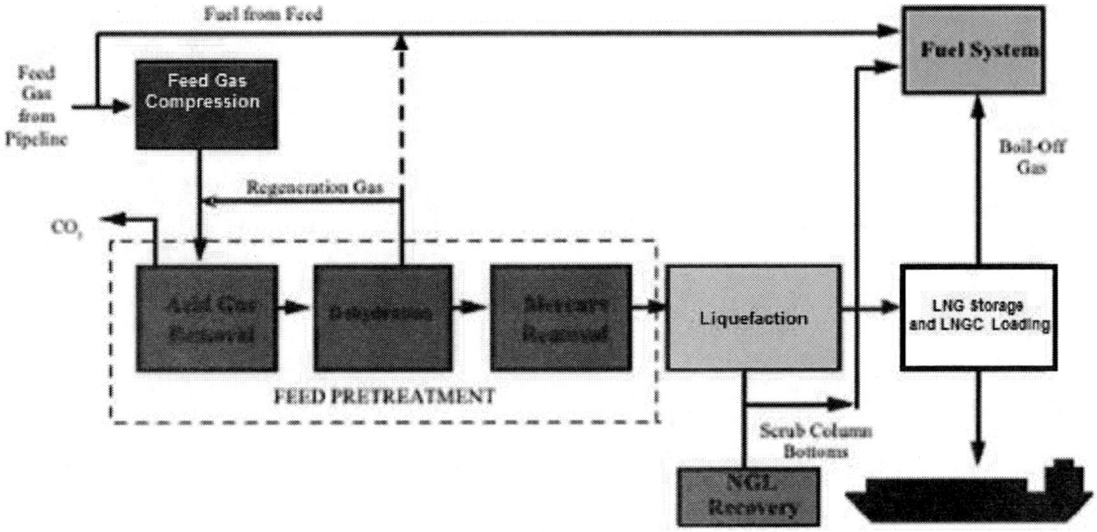

Figure 3. Block Process Flow Diagram. Liquefaction process

Worldwide accidents in LNG plants

The following are major known accidents in LNG facilities in which their design was done in compliance with the applicable standards at that time.

October, 1944 Cleveland, Ohio, USA. LNG Peakshaving Facility
The tanks had been filled to capacity and the cylindrical tank failed; releasing all of its contents into the nearby streets and sewers of Cleveland. The cloud promptly ignited and a fire ensued which engulfed the nearby tanks, residences and commercial establishments. The sphere nearest to the cylindrical tank toppled over and released its contents that immediately evaporated and ignited. In all, 128 people were killed and 225 injured.

The accident was due to the low temperature embitterment of the inner shell of the cylindrical tank. The inner tank was made of 3.5% nickel steel, a material now known to be susceptible to brittle fracture at LNG storage temperature (-165°C (-260°F) approximately). The low nickel content of the tank was due to war rationing of certain metals.

May, 1965 Canvey Island, Essex, United Kingdom. LNG Import Terminal
A small amount of LNG spilled from a tank during maintenance. The spill ignited and one worker was seriously burned. No other details have been made available.

1971 La Spezia, Italy LNG Import Terminal .
This accident was caused by a phenomenon called "rollover," where two layers of LNG having different densities and heat content are allowed to form. The sudden mixing of these two layers results in the release of large volumes of vapour. No ignition took place.

January, 1972 Montreal, Canada. LNG Peakshaving Facility
The accident occurred in the control room due to a back flow of natural gas from the compressor to the nitrogen line. Natural gas entered the control room through the nitrogen header and accumulated in the control room, where operators were allowed to smoke. The explosion occurred while an operator was trying to light a cigarette.

March, 1977 Algeria. LNG Export Facility
A worker at the Camel plant was frozen to death when he was sprayed with LNG, which was escaping from a ruptured valve body on top of an in-ground storage tank. The resulting vapour cloud did not ignite. The valve body that ruptured was constructed of cast aluminium. The current practice is to provide valves in LNG service that are made with stainless steel.

March, 1978 Das Island, United Arab Emirates.LNG Export Facility
A bottom pipe connection of an LNG tank failed resulting in an LNG spill inside the LNG tank containment. The liquid flow was stopped by closing the internal valve designed for just such service. A large vapour cloud resulted and dissipated without ignition. No injuries or fatalities were reported.

October, 1979 Cove Point, Maryland, USA.LNG Import Terminal.
An explosion occurred within an electrical substation. LNG had leaked through an inadequately tightened LNG pump electrical penetration seal, vaporized, passed through 200 feet of underground electrical conduit, and entered the substation. Since natural gas was never expected in this substation, no gas detectors had been installed in the building. The natural gas-air mixture was ignited by the normal arcing contacts of a circuit breaker, resulting in an explosion. The explosion killed one operator in the building, seriously injured a second and caused about $3 million in damages.
Before the Cove Point Terminal was restarted, all pump seal systems were modified to meet the new codes and gas detection systems were added to all buildings.

1989 Thurley, United Kingdom. LNG Peakshaving Facility
While cooling down the vaporizers in preparation for sending out natural gas, low-point drain valves were opened on each vaporizer. One of these drain valves had not been closed when the pumps were started and LNG entered the vaporizers. As a result, LNG was released into the atmosphere as a high-pressure jet. The resulting vapour cloud ignited about thirty seconds after the release began. Two operators received burns to their hands and faces.

January 19, 2004 Skikda, Algeria. LNG Export Facility
A leak in the hydrocarbon refrigerant system formed a vapour cloud that was drawn into the inlet of a steam boiler. The increased fuel to the boiler caused rapidly rising pressure

within a steam drum. The rapidly rising pressure exceeded the capacity of the boiler's safety valve and the steam drum ruptured. The boiler rupture was close enough to the gas leak area to ignite the vapour cloud and produce an explosion due to the confined nature of the gas leak and an ensuing fireball. The explosions and fire destroyed a portion of the LNG plant and caused 27 deaths and injury to 72 more. No one outside the plant was injured nor were the LNG storage tanks damaged by the explosions.

Some of these accidents were caused by bad operation, or during construction work, and not directly related with the design of the plant, but in such cases a better design would definitely mitigate the effects of such accidents.

Safety improvements

In the LNG plants, either regasification or liquefaction, there are many areas with significant amount of gas inventory. LNG storage, refrigerants tanks, equipment... Most of the equipment or storage section have their own safety devices and, according to the different standards and risk analysis, are installed within an exclusion area.

In case of Floating LNG (FLNG) facilities, or Floating Storage and Regasification Units (FSRUs), that have the potential to revolutionize the way natural gas resources are developed by moving the production and processing out to sea, and therefore are relatively new in the industry, the high potential risks and the novelty of the concept require detailed and sometimes new safety studies. Fire and Explosion risk analysis, cryogenic risk analysis, smoke and gas dispersion risk analysis, etc. feed into a Quantitative Risk Assessment (QRA). Performing these safety studies is challenging, given the lack of any feedback and references and in some cases even like of applicable codes.

It could be discussed about increasing safety distances between storage tanks and other equipment to those required by the applicable standards and/or risk analysis, but in this paper we are focusing on the design some depressurization systems in places where there is high pressure or high volume of LNG or NG.

When thinking about an LNG plant, and where the largest inventory of gas is, and therefore where the plant should be safer, the answer is clear: as said, it is the storage section. However, what about the big lines?. There are many pipes in any process plant, and some of the lines in an LNG plant are of significant size. The consideration about safety in the lines is usually limited to the materials to be used, and also insulation requirements, but the applicable standards in today's design do not say much about process safety, besides some recommendations about routing of the lines and or fittings in between or with the equipment. It is in our belief that basic designs of the LNG plants need to consider the process lines in a wider spectrum. This is also significant for offshore LNG facilities, especially for the pipes connecting with the onshore facilities.

Amec Foster Wheeler's Basic Designs of the LNG plants consider this wider view of the process lines.

In case of an incident involving a possible fire in the plant the tanks and the process equipment have their own emergency systems, including pressure release devices.

The emergency blowdown systems are designed to reduce the light hydrocarbon inventory quickly, in order to decrease the duration of an eventual fire in the terminal. These systems can also be designed for those pipes of large size, that vehicle NG.

The depressuring system is designed based on API Standard 521. So the basis for the design is to decrease the pressure of the system to 50% of the design pressure within 15 minutes. Why 15 minutes? 15 minutes is the time that the temperature of a steel plate, that is exposed to a pool fire, will reach about 650°C (1200°F).That's a temperature that is very close to the steel allowable stress condition. When depressing the pipeline within the mentioned 15 minutes to 50% of the initial pressure, then the time to reach that temperature will increase to about 2.5 hours.

The system is composed of an automatic valve and a restriction orifice located downstream the valve. The natural gas from the depressuring system is sent to the flare. In the shown case on Figure 4 the depressing system is located at natural gas high pressure header, downstream the vaporizers outlet and upstream the metering station in a LNG regasification terminal. The high pressure of the natural gas and the large size and length of this line suggest us to provide more safety than that required by the codes.

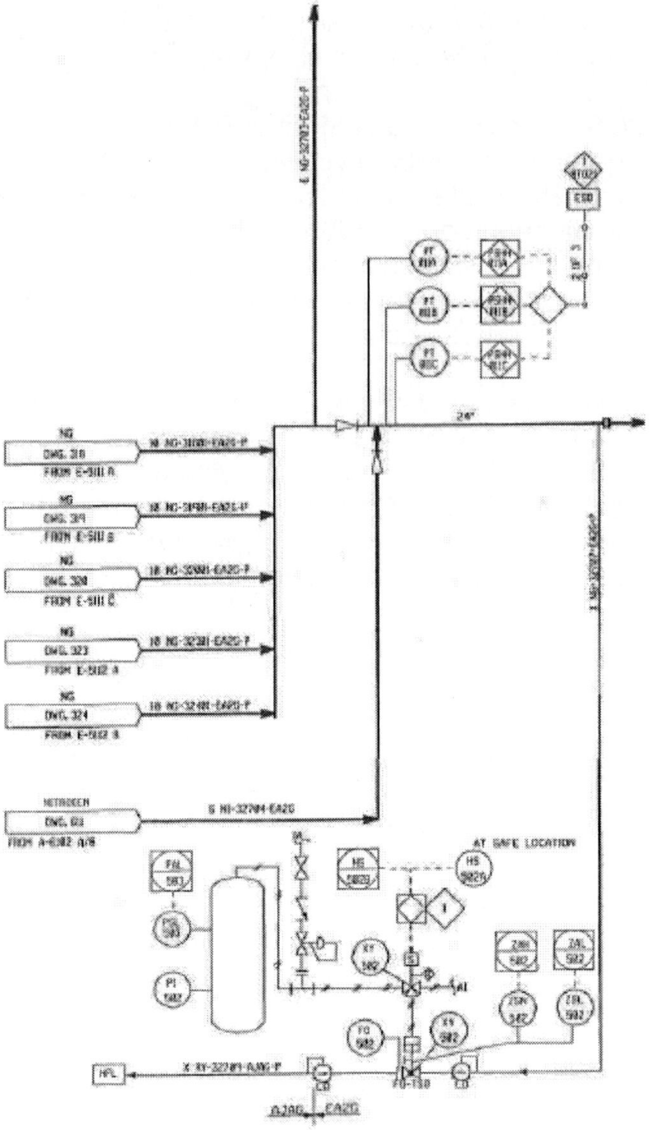

Figure 4. Depressurization system

Neither API, nor NFPA, consider this design. The installation of it does not imply addition to the CAPEX, nor OPEX, of the installation and improves the safety of the plant significantly.

Conclusion

The design of the LNG plants should be done not only to comply the standards but also analyzing all the potential areas that can produce an incident. and add more safety. Also when applying the codes, the design should not be limited to the "at least" minimum requirements. No matter what type of plants we are in, all the improvements in safety should be considered.

Our recommendation is the following:

- Besides following the standards, perform "smart" design to increase safety

- Include depressuring systems in all the large process lines where there is LNG or NG.

Liquid Density by EOS

Yoshitsugi Kikkawa,

Kikkawa Giken, Yokohama Japan,
yoshikikkawa@yahoo.co.jp

David Hill

Manager of Technical Support, Chemstations,
Houston,Texas

davidh@chemstations.com

Liquid Density by EOS

- Liquid Density for Process Engineering
- Equation of State, EOS
- Vapor-Liquid Phase Equilibria by EOS
- Liquid Density by EOS

Liquid Density for Process Engineering

- Liquid Density is Basic Physical Property for Process Engineering
- Hydraulics for Process Piping
- Thermal Rating for Heat Exchanger
  - ➢Main Cryogenic Heat Exchanger
  - ➢Propane Chillers
- Hydraulics for Separator Vessel
- Hydraulics of Tray and Packing for Distillation Column
  - ➢Scrub Column
  - ➢DeC2, DeC3, DeC4
- Hydraulics for Reactor and/or Adsorption Vessels
  - ➢LPG Treater

EOS, Equation of State

- Popular EOS for LNG Plant Design
  - ◆SRK, Soave Redlich-Kwong
  - ◆Peng-Robinson
- SRK and Peng-Robinson has been developed successfully for vapor liquid phase equilibria estimation, sacrificing liquid density estimation of around 8% error
- SFSRK, Scott-Fuller SRK
- VTPR, Volume Translated Peng Robinson

SRK

$$P = \frac{RT}{v-b} - \frac{a(T)}{v(v+b)} \tag{1}$$

Where

$$a(T) = 0.42747\frac{R^2T_c^2}{P_c}\alpha(T) \tag{2}$$

$$b = 0.0867\frac{RT_c}{P_c} \tag{3}$$

Peng-Robinson

$$P = \frac{RT}{V_m - b} - \frac{a(T)}{V_m(V_m+b) + b(V_m - b)} \tag{4}$$

Where

$$a(T_c) = 0.45724\,\frac{(RT_c)^2}{P_c} \tag{5}$$

$$b = 0.07780\frac{(RT_c)}{P_c} \tag{6}$$

The generalized expression for the temperature-dependant parameter is given by

$$a(T) = a(T_c)\alpha(T) \tag{7}$$

where

$$\alpha(T) = \left\{1 + m\left[1 - \sqrt{\frac{T}{T_C}}\right]\right\}^2 \tag{8}$$

With

$$m = 0.3746 + 1.5423\omega - 0.2699\omega^2 \tag{9}$$

SFSRK

Fig. 5.1 Compressibility Factor vs. b/4v
hs: Hard Sphere
Exact: Calculated by Molecular Simulation

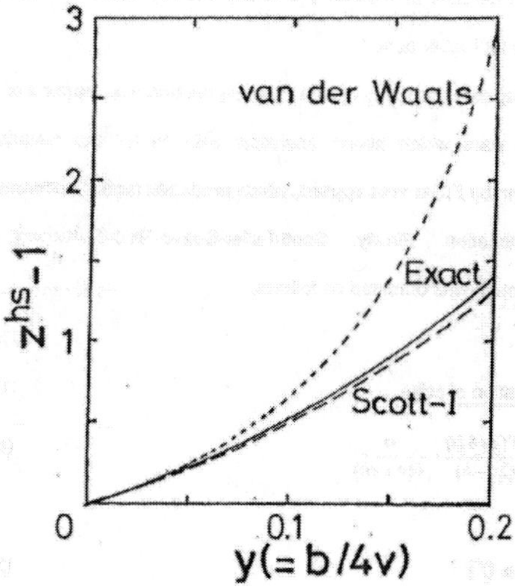

$$p = \frac{RT(2v+b)}{v(2v-b)} - \frac{a}{v(v+cb)} \qquad (5.1)$$

where

$$a_i = a_{ci}\alpha_i(T) \qquad (5.2)$$

$$a_c = \Omega_a(\beta)\frac{(RT_c)^2}{P_c} \qquad (5.3)$$

$$b = \Omega_b(\beta)\frac{RT_c}{P_c} \qquad (5.4)$$

$$c(\beta) = \frac{1}{\beta}\left(\sqrt{\frac{1}{\beta} + \frac{3}{4} - \frac{3}{4}\beta + \frac{1}{8}\beta^2} - \frac{3}{2}\right) \qquad (5.5)$$

and

$$\Omega_b(\beta) = \beta\frac{(4-\beta^2)(2+c\beta) - (4+4\beta-\beta^2)(1+c\beta)}{(2-\beta)^2(2+c\beta)} \qquad (5.6)$$

$$\Omega_a(\beta) = \frac{(4+4\beta-\beta^2)(1+c\beta)^2\Omega_b(\beta)}{\beta(2-\beta)^2(2+c\beta)} \tag{5.7}$$

$$\alpha_i(T) = \{1+q(\beta)(1-T_r^{0.5})\}^2 \tag{5.8}$$

$$q(\beta) = \left(\frac{\beta}{0.345}\right)^{\frac{1}{4}} m \tag{5.9}$$

$$Z_c = \frac{(4-\beta_c^2)(2+c_c\beta_c)-(4+4\beta_c-\beta_c^2)(1+c_c\beta_c)}{(2-\beta_c)^2(2+c_c\beta_c)} \tag{5.10}$$

$$\beta = \beta_c + (\beta_0 - \beta_c)\left(\frac{2}{1+e^{\theta(Tr-1)}} - 1\right) \tag{5.11}$$

$$\beta_c = \frac{b}{v_c} \tag{5.12}$$

$$\theta = 10.9356 + 0.0285P \tag{5.13}$$

VTPR

$$P = \frac{R \cdot T}{(v+c-b)} - \frac{a(T)}{(v+c)\cdot(v+c+b)+b\cdot(v+c-b)} \tag{1}$$

The attractive parameter $a_{ii}(T)$

$$a_{ii}(T) = a_{c,i}\alpha_i(T) = 0.45724\frac{R^2 \cdot T_{c,i}^2}{P_{c,i}} \cdot \alpha_i(T) \tag{2}$$

and the covolume b_{ii}

$$b_{ii} = 0.0778\frac{RT_{c,i}}{P_{c,i}} \tag{3}$$

of the pure component i can directly be calculated from critical data. The temperature dependency of the attractive parameter a_{ii} is taken into account with the help of the Twu α-function[8]

$$\alpha_i(T) = T_{r,i}^{N_i \cdot (M_i - 1)} \cdot \exp[L_i \cdot (1 - T_{r,i}^{N_i \cdot M_i})] \quad (4)$$

which leads, compared to the Mathias–Copeman α-function[9] used in PSRK, to improved results especially at high reduced temperatures.

The parameters L_i, M_i, and N_i of the α-function are usually obtained by fitting eq 4 to experimental pure component vapor pressure data. Since liquid heat capacities also deliver important information about the temperature dependency especially at low temperatures, liquid heat capacities can additionally be taken into account during the parameter-fitting procedure.[10]

The c parameter proposed by Peneloux et al.[11] can be calculated from the mole fractions and the component-specific translation parameters c_i

$$c_i = v_{\mathrm{PR},i} - v_{\exp,i} \ (\text{at } T_{r,i} = 0.7) \quad (5)$$

using a linear mixing rule

$$c = \sum_i x_i c_i \quad (6)$$

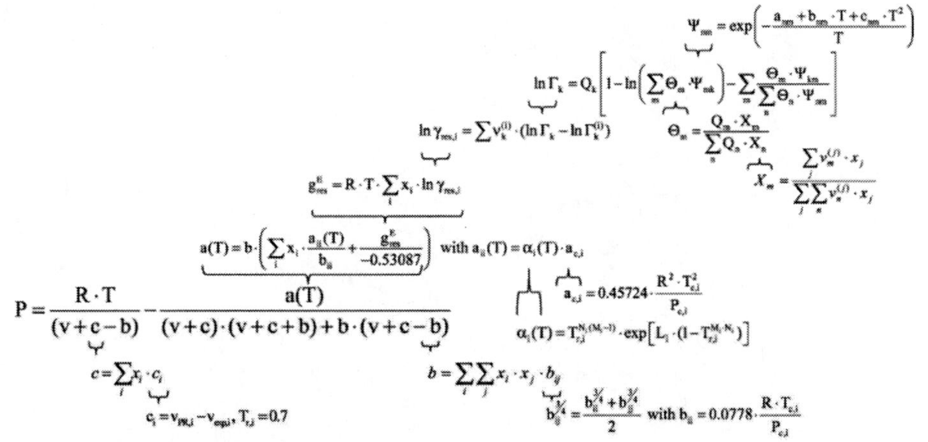

Figure 5. Overview of the VTPR group contribution equation of state.

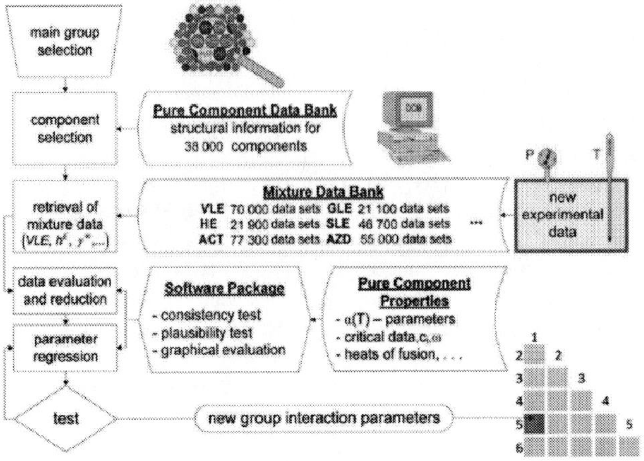

Figure 7. Flow sheet of the group interaction parameter-fitting procedure.

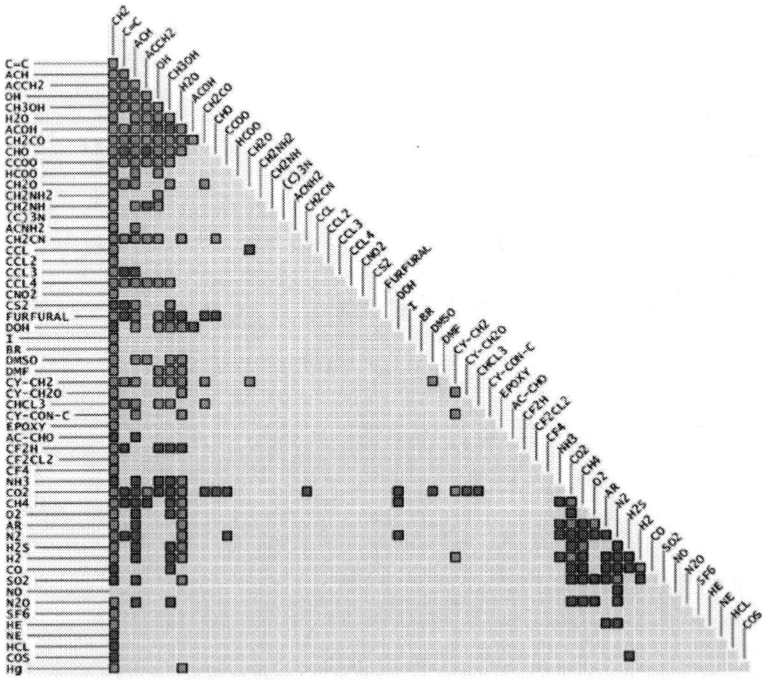

Figure 8. Actual group interaction parameter matrix for VTPR: (blue) two parameters, (green) four parameters, (orange) six parameters.

VTPR Vapor-Liquid Example

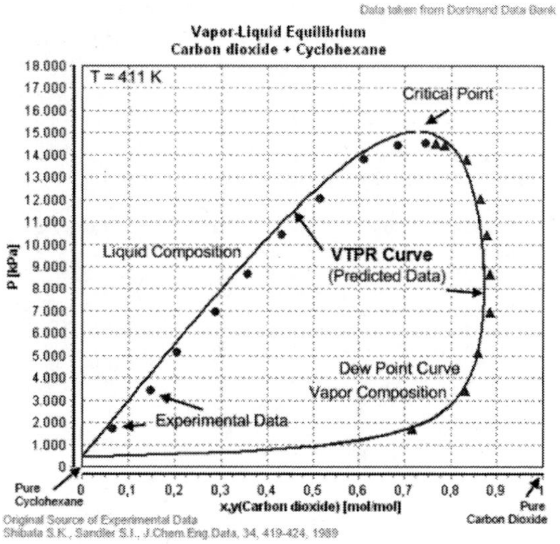

Vapor-Liquid Equilibria(K-Value) by EOS

- Evaluation for SRK and SFSRK
- Selected System
  - ◆ N2+CO2
  - ◆ CH4+CO2
  - ◆ CH4+C3H8

Table 5.4 Correlation Performance of SRK and SFSRK EOS for Vapor-Liquid Equilibria and Saturated Densities

System (1)+(2)	Temp.[K]	EOS	k_{12}	Error[%]* in calculation for x	y	Liq. Density [kg/m³]	Vap. Density [kg/m³]
$N_2 + CO_2$ (Arai et al.[2])	288.15	SRK	-0.006	8.3	9.5	13.2	6.1
		SFSRK	0.180	6.4	7.1	3.6	3.0
	273.15	SRK	0.004	8.1	8.7	12.2	4.7
		SFSRK	0.127	8.9	5.7	3.7	5.2
	253.15	SRK	-0.039	6.7	9.2	8.6	4.1
		SFSRK	0.091	4.1	9.2	4.1	7.3
$CH_4 + CO_2$ (Arai et al.[2])	288.15	SRK	0.139	4.7	4.1	13.8	6.5
		SFSRK	0.181	3.3	8.0	2.1	1.0
	273.15	SRK	0.131	3.2	6.2	10.1	3.7
		SFSRK	0.166	2.2	3.5	3.0	4.8
	253.15	SRK	0.107	4.3	5.1	9.0	2.9
		SFSRK	0.137	4.9	6.0	3.1	7.9
$CH_4 + C_3H_8$ (Reamer et al. [28])	327.59	SRK	0.046	1.3	2.3	12.0	4.0
		SFSRK	0.107	5.3	4.1	1.9	3.7
	310.93	SRK	0.029	5.4	7.1	11.1	3.2
		SFSRK	0.082	4.8	3.3	2.8	5.5
	294.26	SRK	0.020	0.6	1.7	8.6	2.5
		SFSRK	0.050	4.9	1.7	5.5	9.8

$$\text{Error}[\%] = 100 \sum_i \left[\,|\text{Calc.}(i) - \text{Exp.}(i)| / \text{Exp.}(i)\right]/N$$, N= Number of datum points

Liquid Density by EOS

- Evaluation for Pure Component, CH_4, C_2H_6, C_3H_8
- Comparison against NIST Data
- EOS
  - ◆SRK
  - ◆Peng-Robinson
  - ◆SFSRK
  - ◆VTPR

CH_4 Density

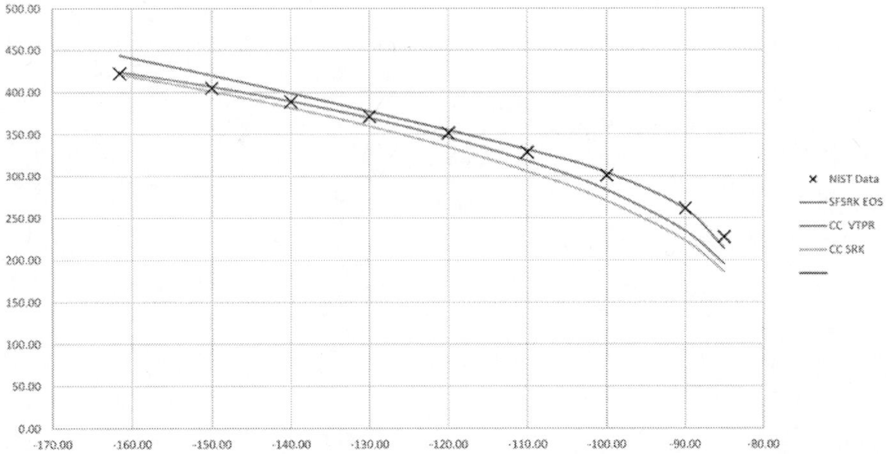

C2H6 Density

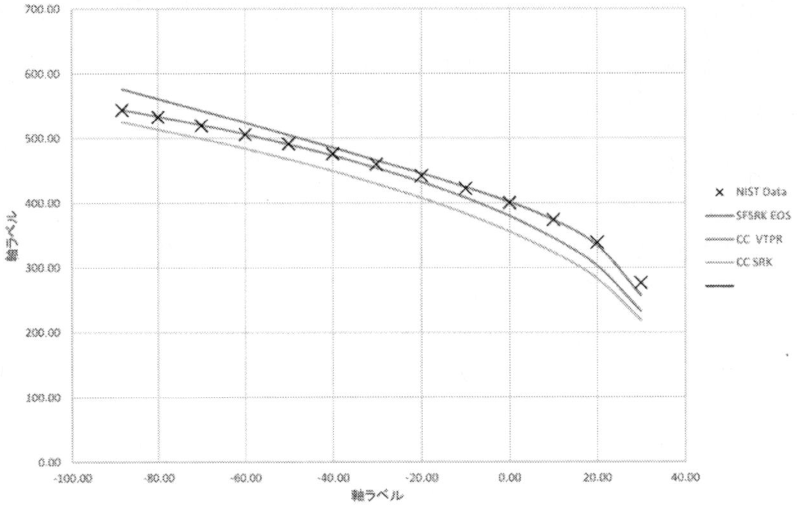

C3H8 Density

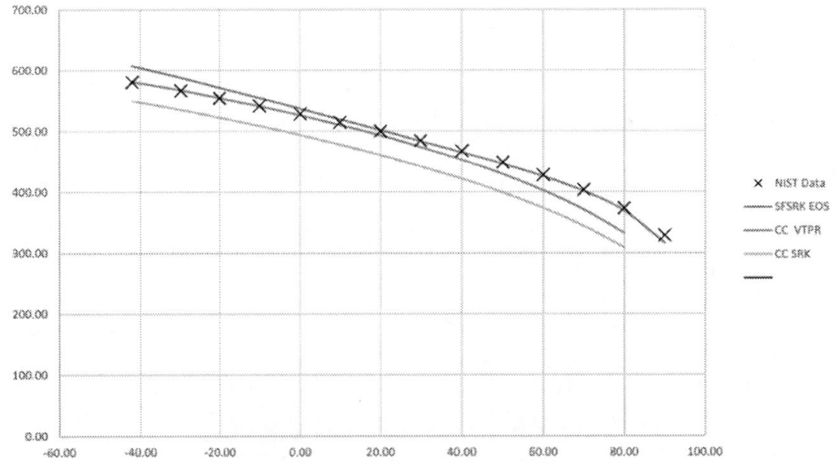

Thank You
for
Your Kind Attention

170b

Versatile Cryogenic Nitrogen Rejection

Heinz C. Bauer (speaker)

Linde Engineering, Pullach, Germany

Abstract

Nitrogen is considered as ballast in natural gas and other fuel gas streams. Thus, most of the pipeline gas specifications limit the nitrogen content to 3 mol%. Fuel gas quality for gas turbines imposes other nitrogen constraints with respect to stability of Wobbe index. Further, the nitrogen concentration in LNG shall not exceed 1 mol% as otherwise stratification of LNG with varying density over the height of the tank might cause the dangerous roll-over effect, which can cause large methane vapor releases or even damage the integrity of flat bottom tanks.

Sophisticated nitrogen rejection unit (NRU) concepts will be discussed, which have been successfully applied to extract nitrogen from gaseous or liquid hydrocarbon fluids and to clean nitrogen for further use or for safe immission into the atmosphere. These case studies include

- N2 rejection from end-flash gas of a LNG plant
- Processing of N2 rich boil-off gas into LNG and fuel
- Clean-up of pipeline gas
- Handling of natural gas or associated gas with increasing N2 content
- CO2 tolerant operation of a NRU

Linde Engineering, which is active in the field of cryogenic methane/nitrogen separation since 40 years provides detailed insights into process technology and corresponding equipment as well as operational experience.

Introduction

Nitrogen is rejected only, if this cannot be avoided. In contrast to other natural gas processes like ethane or LPG extraction (also known as NGL extraction) the handled component 'nitrogen' is of no commercial value. Thus, nitrogen rejection is a cleaning step like sour gas removal. The added value lies in more attractive properties of the treated feed stock like

- increased Wobbe Index of the processed natural gas or
- reduced risk of LNG stratification in a flat bottom tank

Having said this, it is obvious that only highly customized concepts for nitrogen rejection will contribute to a commercially viable and successful project. None of the plants, which will be discussed in this paper, are identical as Basis of Design and clients' expectations are too different to allow for a meaningful standardization of the solutions.

Flash Processes

Nitrogen shows a significantly higher volatility than methane. At atmospheric conditions the molar ratio N_2/CH_4 is about 20 times higher in the vapor phase than in the corresponding liquid phase. This large 'k-value' between nitrogen and methane provides an elegant means to design a bulk removal of nitrogen from LNG.

Most of the modern LNG plants include an 'end-flash' system between the pressurized liquefaction section and the flat bottom LNG storage tanks. If the inlet temperature to this 'end-flash' is selected properly the resulting liquid will meet the requirement of max 1 mol% nitrogen, while the flashed-off gas will show a double digit percentage of nitrogen.

This simple and well proven arrangement, however, finds its limits, if the feed gas is too rich in nitrogen. Starting from around 3 mol% nitrogen in the feedstock either the flow rate or the nitrogen content of the flash gas will be too high for easy utilization.

Advanced process arrangement

Boil-off gas (BOG) reliquefaction is a special case of nitrogen rejection from light hydrocarbons as BOG frequently contains high levels of nitrogen in the range of 4-20 mol%. As the reliquefied LNG shall adhere to the same strict nitrogen limit of <1 mol% a cost effective and robust process is desirable, which avoids complex operation of cryogenic separation columns.

Figure 1 shows the core components of an efficient BOGR process. BOG is compressed first to an elevated pressure preferably above its critical pressure, cooled down to below the critical temperature of the BOG (typically <-100°C) by external refrigeration of any type and flashed into a vessel, which is operated at intermediate pressure. The gas from this IP drum is well suited as fuel gas for gas turbines with high nitrogen tolerance. The liquid from the IP drum is further flashed into a LP drum, which produces on-spec LNG with <1 mol% nitrogen. The flash gas from the LP drum is recycled to the feed stream.

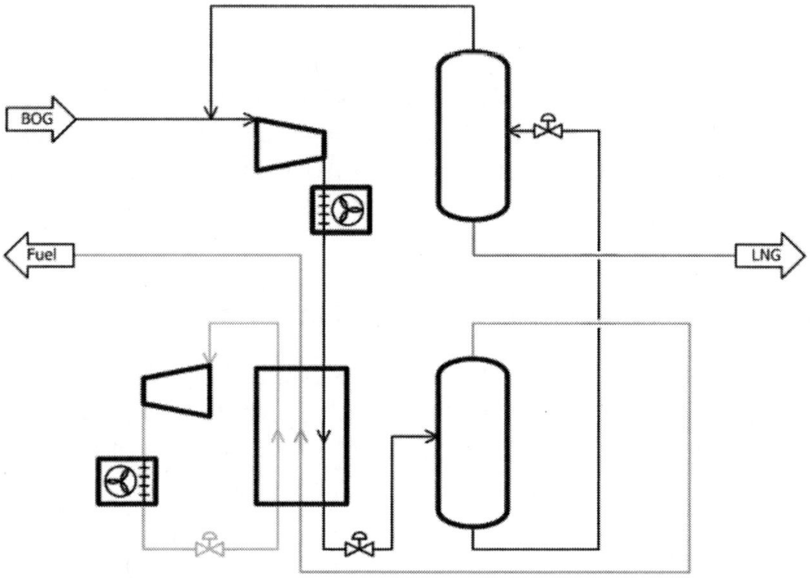

Figure 1 Refrigerated Double Flash Process

Selected references

The MARLIN (**M**LNG **A**dvanced **R**e-**LI**quefactio**N**) project (see Figure 2) of Malaysia LNG Sdn. Bhd. in Bintulu is an excellent example for a plant designed on the basis of the Refrigerated Double Flash process. It has been started up in 2015 and has passed the performance test successfully.

Figure 2 Boil-Off Gas Reliquefaction Plant, Malaysia LNG, Bintulu

Single Column Processes

Flash processes are robust and easy to operate, but cannot produce a nitrogen rich fraction, which can be sent directly to the atmosphere as the emission of methane is limited in most developed countries. Hence, a nitrogen purity of at least 99 mol% is requested. Still non-cryogenic alternatives based on membrane or adsorption techniques are not competitive for full scale applications above 20,000 Nm³/h (20 MMSCFD) of feed gas. Therefore, the use of cryogenic distillation columns as they are known from air separation remains the best tool to deploy.

A widespread arrangement uses only one single column, which is operated at 25-30 bar so that technically pure nitrogen can be condensed in the top part against a fraction of the methane rich bottoms vaporizing at low pressure. This concept has its merits for low-to-medium nitrogen content (3-50 mol%) in the feedstock, but suffers from high compression power for the recompression of the methane fraction that had been used in the overhead condenser. Further, the operability of this process depends strongly on the composition of the bottom product. High C_{2plus} content will reduce the dew point pressure such that undesirable vacuum conditions may occur on the suction side of the compressor. Even more concerning is a poor tolerance towards CO_2 as already 100 vppm of CO_2 in the bottoms fraction will cause severe solid precipitation (icing) in the overhead condenser.

Advanced process arrangement

Figure 3 shows an answer to the challenges mentioned above. Compression power can be reduced significantly using more than one pressure level in the refrigeration loop, which is rather a heat pump loop to stay technically correct. A side condenser for the N_2/CH_4 column is well invested money to reduce OPEX.

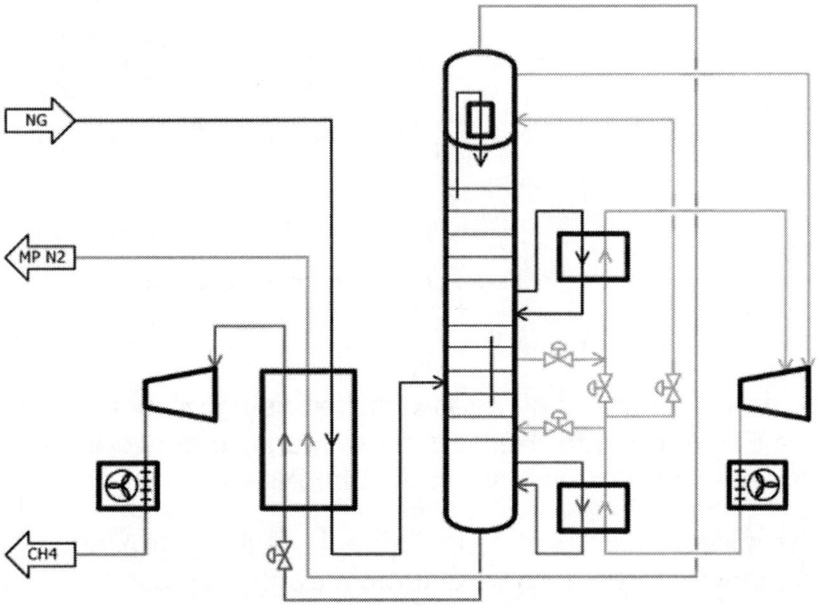

Figure 3 Heat Pumped & CO2 Tolerant Single Column Process

Another outstanding feature is a special tap point for the refrigerant make-up. Instead of using the potentially contaminated (CO_2, C_{2plus}) bottoms of the column a side draw 'behind' a column partition offers the perfect refrigerant. Adjusting the column traffic properly will push all disturbing components on the feed side of the partition to the bottoms without polluting the make-up stream on the draw side of the partition.

Selected references

The AMUR Gas Processing Plant (GPP) close to Belogorsk (see ellipse in Figure 4) is a very important part of Gazprom's Eastern Gas Program. Linde Engineering will supply engineering services and equipment for one of the world's largest natural gas processing plants including C_{2plus} extraction and fractionation, nitrogen rejection based on the Heat Pumped & CO_2 Tolerant Single Column process as well as helium purification and liquefaction.

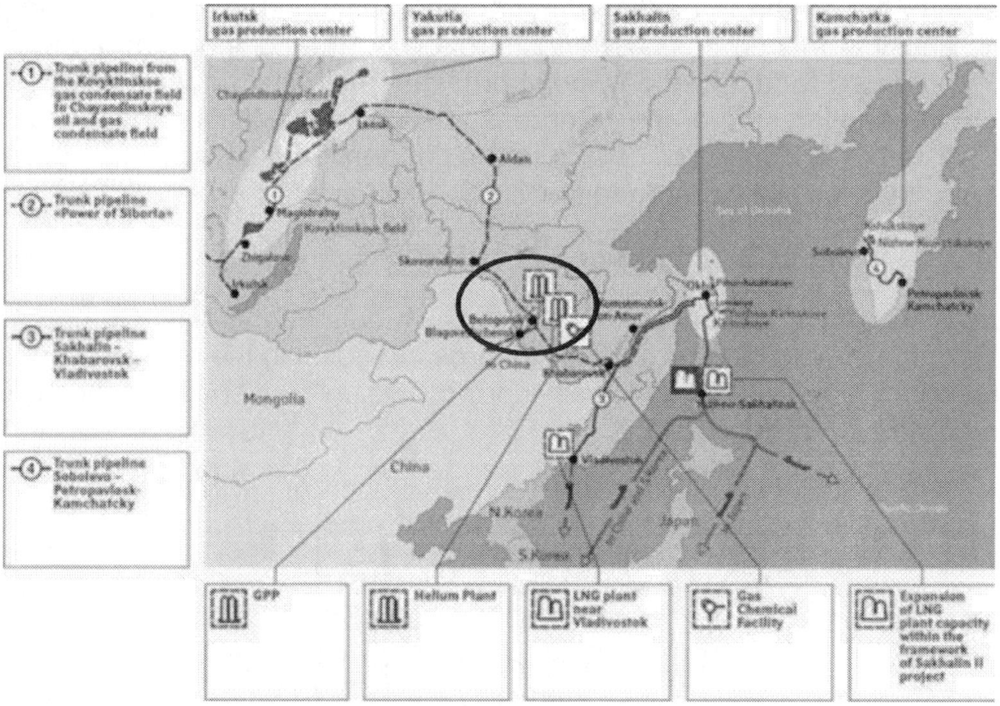

Figure 4 AMUR GPP as part of the Eastern Gas Program (source Gazprom)

Another reference plant for nitrogen rejection using a single column process has been executed by Linde Engineering for Helison, a joint venture between Linde and Sonatrach for a helium purification and liquefaction plant in Skikda, Algeria.

Further, the mid-scale LNG plant in JiMunai, Xinjiang province, P.R. China owned and operated by Guanghui features a single column nitrogen rejection step, which is integrated into the liquefaction scheme.

Double Column Processes

Single column concepts produce a nitrogen fraction under pressure. This may be of special interest, if this fraction contains a relevant concentration (>3 mol%) of helium as its purification and liquefaction benefits from the elevated pressure level. In many cases, however, the rejected nitrogen is sent to atmosphere and the energy of the pressurized nitrogen is lost. Air separation technology teaches a clever column configuration, where the overhead of the column is not refrigerated with boiling heavies (oxygen in case of an ASU, methane in case of a NRU), but with boiling nitrogen. In this 'double column' process the overhead of the lower (high pressure) column is condensed against boiling bottoms of the upper (low pressure) column. Here, vaporizing the methane product is not required to operate the rectification.

The heat integration of a double column system is quite sophisticated. To generate enough reflux for the low pressure column with the correct composition at least 25 mol% nitrogen has to be present in the feed gas. In case the nitrogen content is too low an enrichment step will be installed upstream of the double column. This could be a simple separator for the partially condensed feed gas or – more efficient – just another column, which pushes the nitrogen concentration in the feed stock to the double column above the mentioned threshold.

Advanced process arrangement

As indicated above there is no need to vaporize the methane fraction to keep the separation running. However, the methane fraction is vaporized in most of the applications to cover the cooling demand of the process, which is driven by J-T refrigeration only. In case surplus refrigeration is readily available e.g. from a LNG plant, the methane fraction can be sent to BL in liquid form. This concept (see Figure 5) eliminates the need for a methane product compressor.

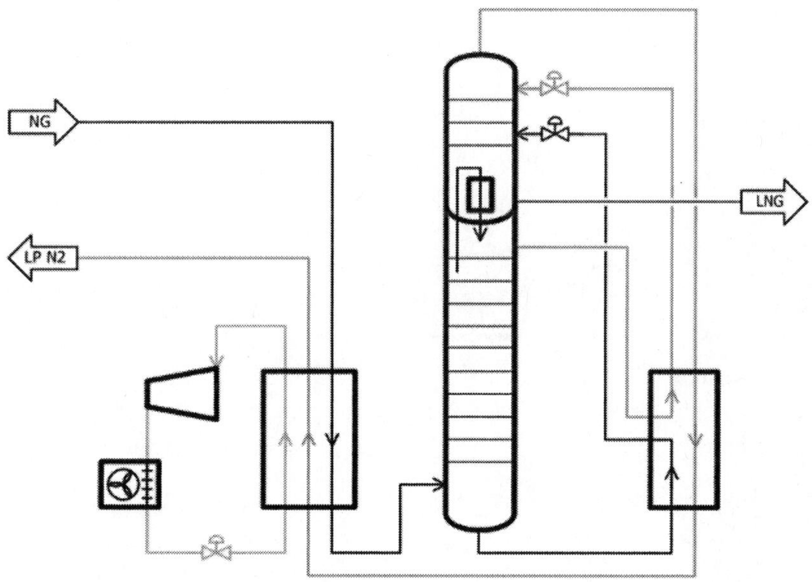

Figure 5 Refrigerated Double Column Process

Selected references

The largest European LNG plant (Snøhvit LNG), which is located in Hammerfest, Norway and is operated by Statoil is equipped with a NRU (see white ellipse in Figure 6) based on the patented Refrigerated Double Column process. The actual methane content in the vented nitrogen is well below 100 vppm.

Figure 6 Refrigerated Double Column Process

A central facility within Woodside's Pluto LNG complex near Karratha, Australia is a Linde NRU (see Figure 7) based on a double column process. The NRU acts as central hub within the fuel gas network providing different gas qualities to the various consumers.

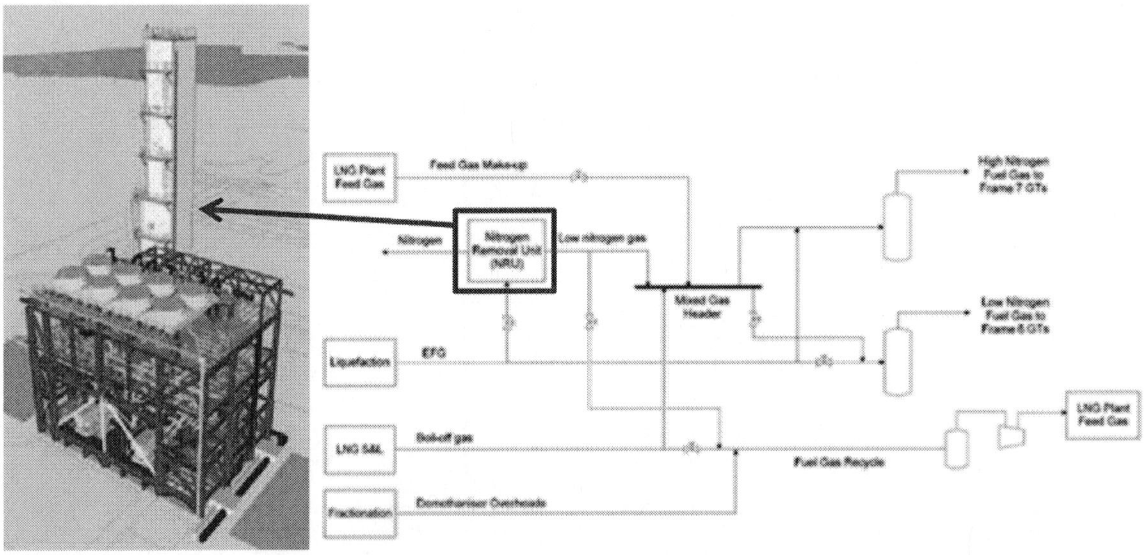

Figure 7 Double Column Process as central hub within the fuel gas network

Conclusions

Linde Engineering has a proven and comprehensive expertise in customizing nitrogen rejection plants to a wide range of applications. This expertise is an excellent platform for future challenges in nitrogen rejection such as:

- Nitrogen based enhanced oil recovery (EOR) projects
- Unlocking large nitrogen rich natural gas reservoirs (e.g. Natuna)
- Tie-back of nitrogen rich reservoirs into existing LNG plants

Overall Reference List

Start-up	Client	Location		Feed Gas		Product(s)		Process
2021	Gazprom	Svobodny	Russia	6x 815,000	Nm³/h	6x 732,000 6x 55,000 6x 28,000	Nm³/h sales gas Nm³/h C_{2+} Nm³/h N_2/He	Heat Pumped & CO_2 Tolerant Single Column
2017	not to be disclosed	-	Australia	255,000	Nm³/h	246,000	Nm³/h sales gas	Double column + enrichment col
2015	Malaysia LNG	Bintulu	Malaysia	130,000	Nm³/h	2,100 400	tpd LNG tpd fuel gas	Refrigerated double flash
2012	Woodside Burrup	Karratha	Australia	78,000	Nm³/h	36,000	Nm³/h fuel gas	Double column
2011	Guanghui	Jimunai	P.R. China	62,500	Nm³/h	1,200	tpd LNG	Single column
2007	Snøhvit Group	Hammer-fest	Norway	71,400	Nm³/h	53,000	Nm³/h	Refrigerated double column
2005	Helison	Skikda	Algeria	47,000	Nm³/h	4,700 2,200 40,000	Nm³/h N_2 Nm³/h He Nm³/h sales gas	Single column
1998	Warren Petroleum		USA	11,200-47,400	Nm³/h		pipeline quality NGL	Double column
1998	NGC/Warren Petro-leum	Hender-son Coun-ty, TX	USA	55,800	Nm³/h			Double column
1994	BHP Petroleum	Onslow	Australia	47,000	Nm³/h	40,000 22,000	Nm³/h sales gas, t/a propane	Single column
1988	Sun Ex-ploration & Production		USA	20,100	Nm³/h		pipeline quality sales gas, Y-grade NGL	Double column
1983	Phillips Petroleum	Dumas	USA	67,000	Nm³/h	160,000 3,200	t/a NGL, t/a crude He	Double column
1982	GB&E	Clenze	Germany	100,000	Nm³/h	36,500	Nm³/h methane	Double column
1977	GB&E	Clenze	Germany	100,000	Nm³/h	36,500	Nm³/h methane	Double column

17 AIChE Spring Meeting, San Antonio, TX — 13TH GLOBAL CONGRESS ON PROCESS SAFETY

Automatic Cooldown of a Coil Wound Heat Exchanger

Dr. Matthew Okasinski, Dr. Fei Chen, Ted Sabram, Brian Johnson and Mark Roberts

LNG Process Technology, **Air Products and Chemicals**

AIChE Spring Meeting
29 March 2017

AIR PRODUCTS

17 AIChE Spring Meeting, San Antonio, TX — 13TH GLOBAL CONGRESS ON PROCESS SAFETY

AP-C3MR™ Process Schematic

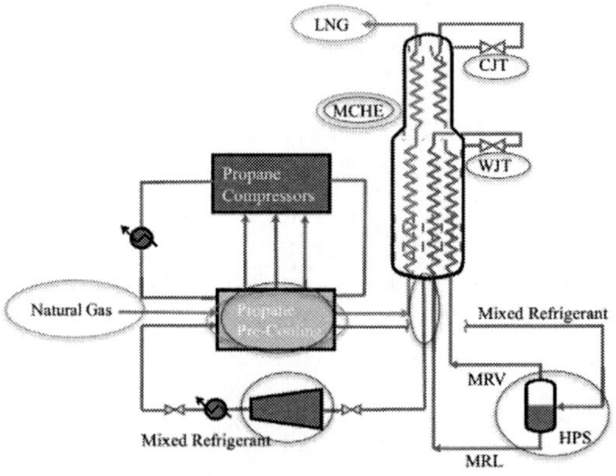

17 AIChE Spring Meeting, San Antonio, TX — 13TH GLOBAL CONGRESS ON PROCESS SAFETY

Key Valves

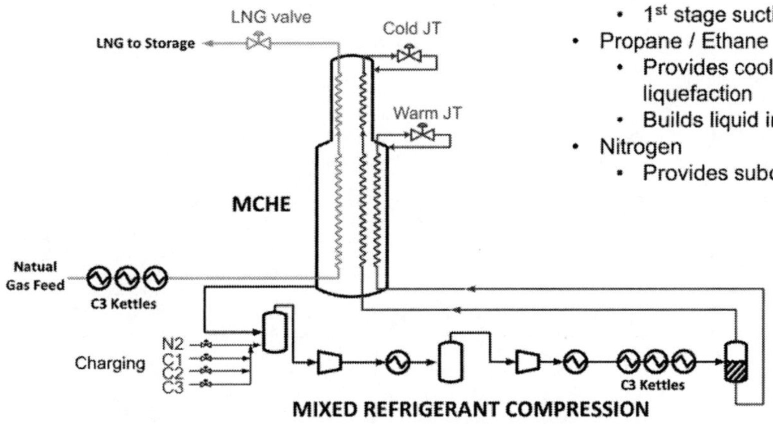

- Methane
  - 1st stage suction pressure
- Propane / Ethane
  - Provides cooling and liquefaction
  - Builds liquid inventory
- Nitrogen
  - Provides subcooling

17 AIChE Spring Meeting, San Antonio, TX — 13TH GLOBAL CONGRESS ON PROCESS SAFETY

MCHE Cooldown

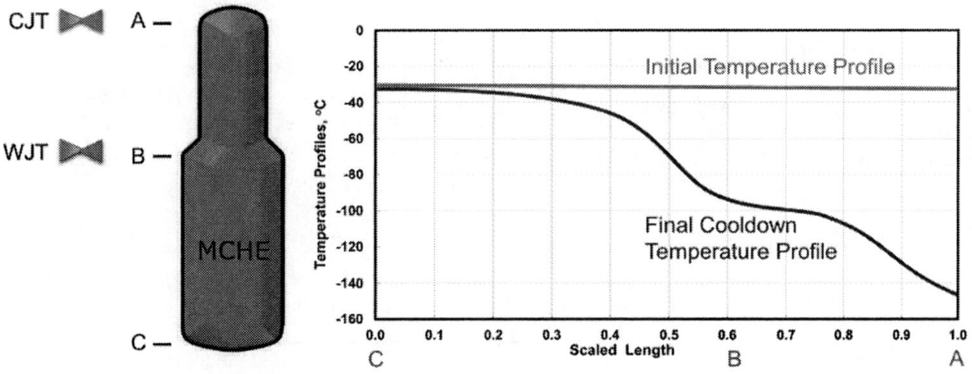

17 AIChE Spring Meeting, San Antonio, TX
13TH GLOBAL CONGRESS ON PROCESS SAFETY

Manual Cooldown

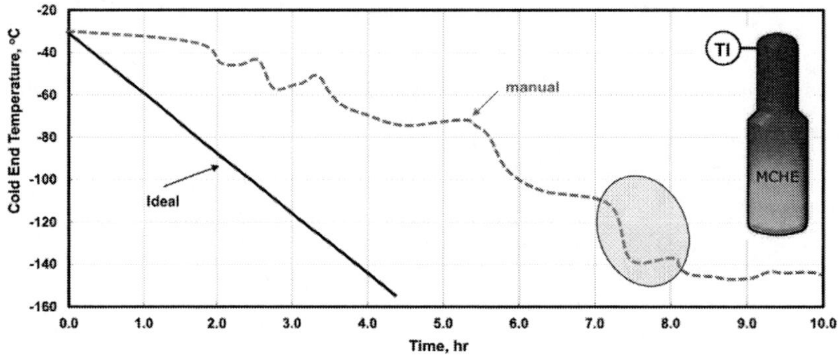

17 AIChE Spring Meeting, San Antonio, TX
13TH GLOBAL CONGRESS ON PROCESS SAFETY

Moderating Cooldown Rate

- As MR starts to condense, cooldown rate will increase
- Proper adjustment of MR and feed flows along with MR composition moderates the cooldown rate.

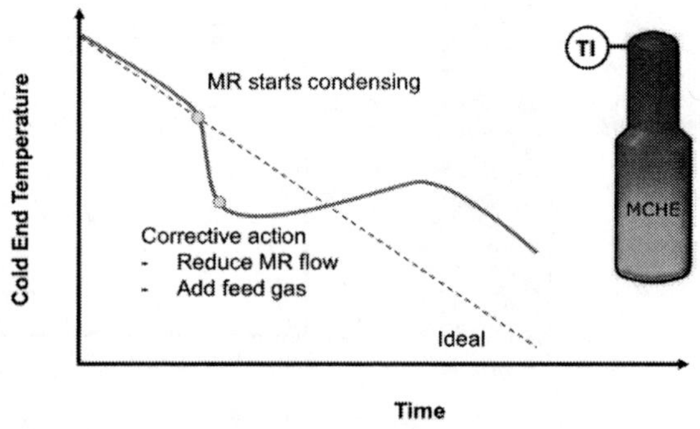

Automated Cooldown Objectives

- Maintain a smooth cooldown at predefined rate
- Minimize cold end flaring
- Consistent cooldowns

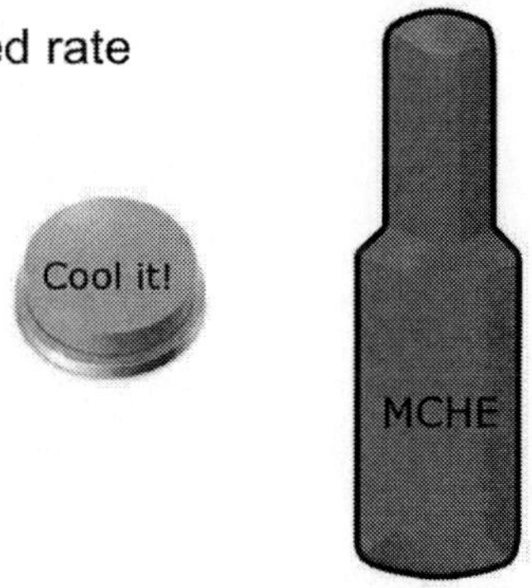

How Automated Cooldown Works

- Continuously ramp up refrigeration during the cooldown
- Initially most refrigeration cools down MCHE metal
- Absorb excess refrigeration by adjusting feed
- With time, more refrigeration goes to cooling feed gas than cooling metal

Automated Cooldown

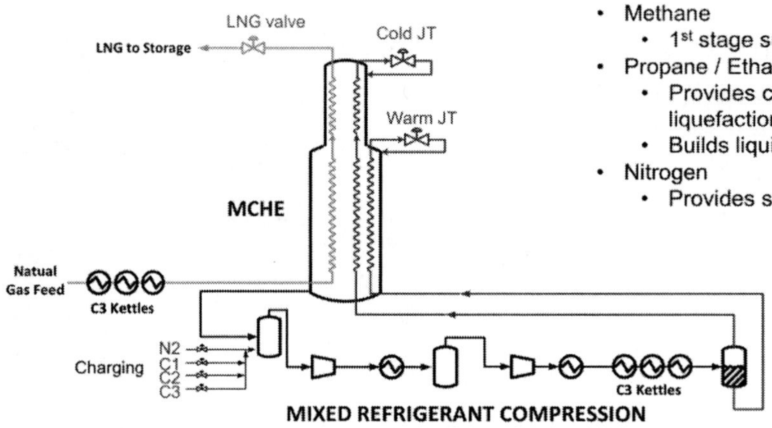

- Methane
  - 1st stage suction pressure
- Propane / Ethane
  - Provides cooling and liquefaction
  - Builds liquid inventory
- Nitrogen
  - Provides subcooling

Applying Constraints

- Constraints maintain process within acceptable conditions
- Example: High level in HP MR Separator will stop make up of C2/C3

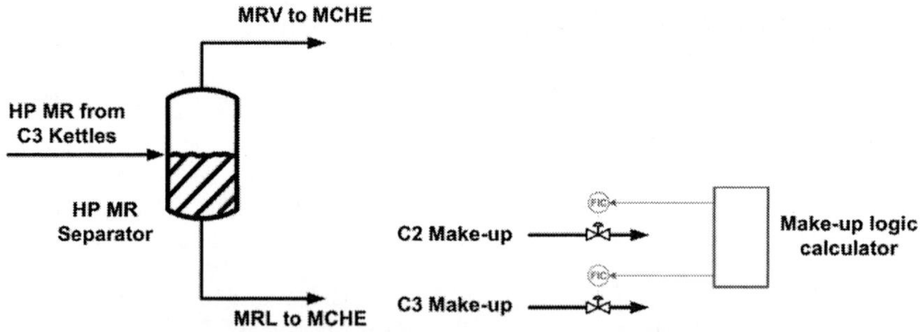

17 AIChE Spring Meeting, San Antonio, TX — 13TH GLOBAL CONGRESS ON PROCESS SAFETY

Dynamic Simulation

- Air Products proprietary models
- Input
  - Rigorous first-principle models
  - System inventory and volumes
  - MCHE geometry and compressor curves
  - Desired cooldown rate and ramping parameters
- Output
  - MCHE temperature profile
  - Cooldown rate

17 AIChE Spring Meeting, San Antonio, TX — 13TH GLOBAL CONGRESS ON PROCESS SAFETY

Various Cooldown Scenarios

	Cold Restart	Warm Restart	Initial Warm Start
MCHE cold end Temperature	Not cold enough	Warm	Warm
MCHE warm end temperature	Normal to cold	Normal	Normal
MR inventory	Normal	Normal	Low
MR separator levels	Established	Established	None

Initial Warm Start – Automated Temperature Profiles

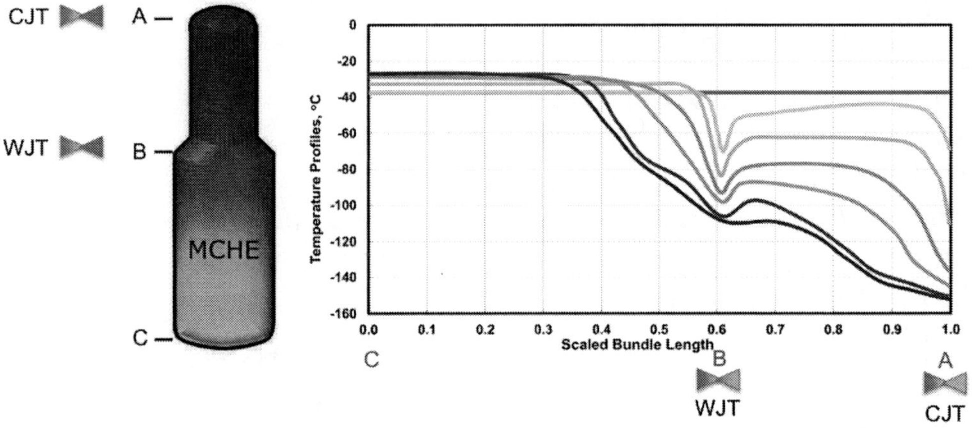

Initial Warm Start – Cooldown Rate

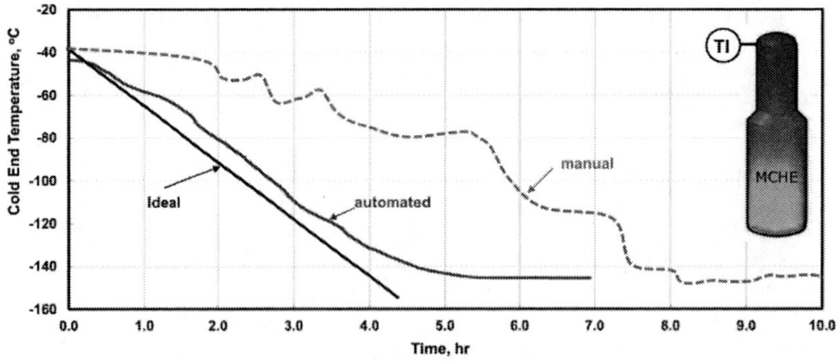

Introducing AP-AutoCool™ Program

- Automated Cooldown Method
  - Software implemented in the plant's DCS system
- Developed from Air Products' experience
  - LNG plant operation and start-ups
  - Dynamic simulation
- Proprietary and patent pending
- Conventional controller and interlock logic setups
- Can be applied to other Air Products' liquefaction processes

Benefits of AP-AutoCool™ Program

- Makes MCHE cooldown operation
  - Faster than manual
  - Smoother than manual
  - Consistent
- Potentially increases
  - annual production
  - plant availability
- Reduces
  - cold end flaring
  - operator intervention
- Suitable for new and existing facilities
  - Handles various cooldown scenarios
  - No additional hardware needed to implement
  - Customizable

AP-AutoCool™ - *fast, consistent, and versatile*

17 AIChE Spring Meeting, San Antonio, TX

13TH GLOBAL CONGRESS ON PROCESS SAFETY

Thank you

tell me more

AIR PRODUCTS

2017 AIChE® SPRING MEETING & 13TH GLOBAL CONGRESS ON PROCESS SAFETY

Anatomy of a PT Flash

Anand Solanki, PE
Sr. Process Engineer
adsolank@bechtel.com

Bechtel Oil, Gas & Chemical, Houston, TX

American Institute of Chemical Engineers
2017 Spring Meeting - 13th Global Congress on Process Safety
March 29, 2017, San Antonio, Texas, USA

17AIChE
Spring Meeting, San Antonio, TX

Agenda

- Basic phase equilibrium equations
- Stability analysis to verify the solution
- Implementation using Successive substitution
- Demonstration

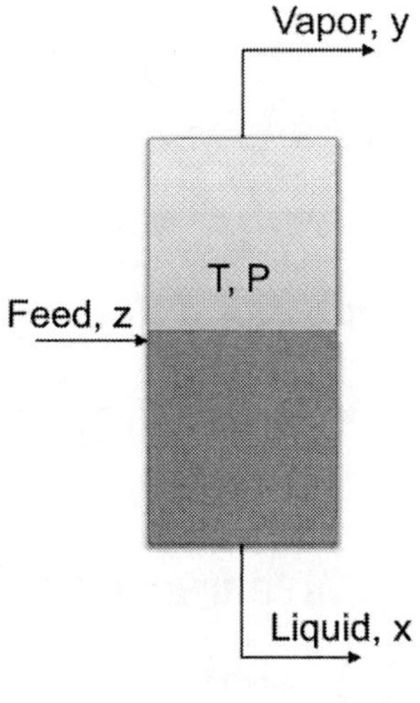

17AIChE
Spring Meeting, San Antonio, TX

2-Phase P-T flash

- Most important equilibrium calculation
- Based on ideal solution behavior
- Successive substitution as Classical method for solving
- Consider two-phase equilibrium in N components mixture with overall composition z.

17 AIChE Spring Meeting, San Antonio, TX

BECHTEL

Basic Phase Equilibrium Equations - 1

- Fugacities of the individual components have the same value in the two phases
  - $f_i^l = f_i^v$, where i = 1, 2, …N **(Eq. 1)**
- Material balance for each component
  - $\beta y_i + (1-\beta)x_i = z_i$ **(Eq. 2)**

    - Where β is overall vapor fraction
- Mole fraction relationship
  - $\sum_{i=1}^{N}(x_i) = 1 \qquad \sum_{i=1}^{N}(y_i) = 1$
  - $\sum_{i=1}^{N}(y_i - x_i) = 0$ **(Eq. 3)**

17 AIChE Spring Meeting, San Antonio, TX

BECHTEL

Basic Phase Equilibrium Equations - 2

- Equation 1 and equation 2 each yields N relationship
- Total of 2N + 1 relationships between 2N + 3 unknown: x_i, y_i, P, T, and β
- Two Specifications are needed
- Type of specifications determines the appropriate solution procedures
  - (P, T), (P, H), (P,S) or (T, V)
  - (T, β), (P, β)

17 AIChE Spring Meeting, San Antonio, TX

BECHTEL

Stability Analysis

- In EOS (Equation of State) methods
  - Solution with liquid and vapor phase of identical properties and identical composition automatically satisfies basic phase equilibrium equations – may cause problem for critical region
- Solution needs verification – Stability method
  - Solution is stable if and only if Gibbs energy at current (T, P) is at its global minimum

17 AIChE Spring Meeting, San Antonio, TX

BECHTEL

Successive Substitution - 1

- Equilibrium Factors:
  - $K_i = \frac{y_i}{x_i} = \frac{\varphi^l_i}{\varphi^v_i}$ **(Eq. 4)**
- Substituting Eq. 4 in material balance equation 2 yields:
  - $x_i = \frac{z_i}{1-\beta+\beta K_i}$ $y_i = \frac{K_i z_i}{1-\beta+\beta K_i}$ **(Eq. 5)**
- Rachford-Rice Equation
  - $\sum_{i=1}^{N}(y_i - xi) = \sum_{i=1}^{N}\left(\frac{z_i(K_i-1)}{1-\beta+\beta K_i}\right) = 0$ **(Eq.6)**

17 AIChE Spring Meeting, San Antonio, TX

BECHTEL

Successive Substitution - 2

- $f(\beta) = \sum_{i=1}^{N}\left(\frac{z_i(K_i-1)}{1-\beta+\beta K_i}\right) = 0$ **(Eq.7)**
- Solving Eq. 7 using Newton's method:
  - $f'(\beta) = -\sum_{i=1}^{N}\left(z_i\frac{(K_i-1)^2}{(1-\beta+\beta K_i)^2}\right) < 0$
- For β = 0 and 1 function "f":
  - $f(0) = \sum_{i=1}^{N} z_i K_i - 1 > 0$ **(Eq. 8)**
  - $f(1) = 1 - \sum_{i=1}^{N} z_i/K < 0$ **(Eq. 9)**

17 AIChE Spring Meeting, San Antonio, TX

BECHTEL

Successive Substitution - 3

- Component i for which K > 1, yields $y_i \leq 1$ from mole fraction relationship
  - $y_i = \frac{K_i z_i}{1-\beta+\beta K_i} \Rightarrow \quad 1-\beta+\beta K_i \geq K_i z_i$
    - $\beta \geq \frac{K_i z_i - 1}{K_i - 1}$ **Eq. (10)**
- Component i for which K < 1, yields
  - $x_i = \frac{z_i}{1-\beta+\beta K_i} \Rightarrow \quad 1-\beta+\beta K_i \geq z_i$
  - $\beta \leq \frac{1-z_i}{1-K_i}$ **Eq. (11)**

17AIChE Spring Meeting, San Antonio, TX

BECHTEL

Successive Substitution - 4

- Check Equation 8, 9 for existence, terminate if both are not satisfied
- Using equation 10 and 11 check if tighter bound can be set for β_{min} and β_{max}
- Calculate β_{ini} by taking average of β_{min} and β_{max}
- If $f(\beta_{ini}) > 0$ then $\beta_{min} = \beta_{ini}$ else, $\beta_{max} = \beta_{ini}$

17AIChE Spring Meeting, San Antonio, TX

BECHTEL

Basic Algorithm

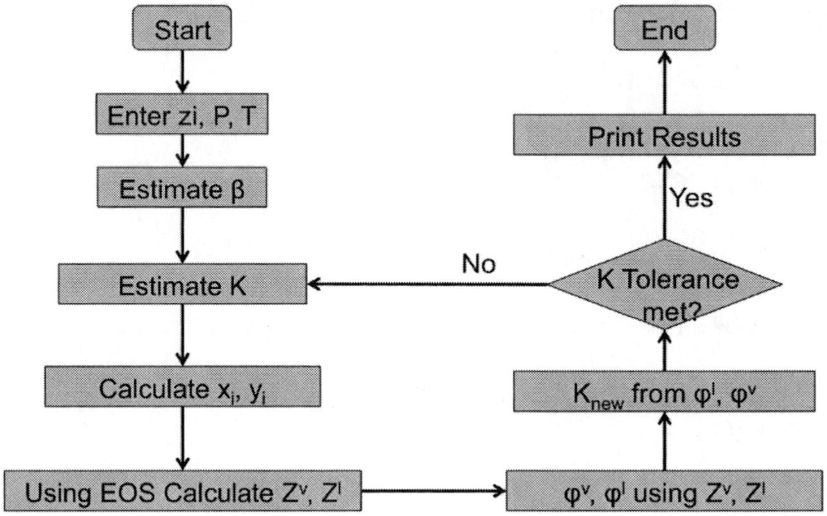

17 AIChE Spring Meeting, San Antonio, TX

BECHTEL

References

- Muskat, M., & McDowell, J. M. (1949, November 1). An Electrical Computer for Solving Phase Equilibrium Problems. Society of Petroleum Engineers. doi:10.2118/949291-G
- Rachford, H. H., & Rice, J. D. (1952, October 1). Procedure for Use of Electronic Digital Computers in Calculating Flash Vaporization Hydrocarbon Equilibrium. Society of Petroleum Engineers. doi:10.2118/952327-G
- M.L. Michelsen, (1993). Phase Equilibrium Calculations. What is easy and what is difficult? Computers Chem. Eng., Vol 17, No 5/6, pp. 431-439
- M.L. Michelsen, (1982).The Isothermal Flash Problem, Part 1-Stability, Fluid Phase Equilibria, 9: 1-19
- Stanley I, Sandler. Chemical and Engineering Thermodynamics

17 AIChE Spring Meeting, San Antonio, TX

BECHTEL

480077 Dynamic Analysis of Feed Gas Control in LNG Plants

Abstract:

In an inlet facility of a natural gas plant, the feed gas or mixture of gas and condensate are received by means of high pressure pipelines. The condensate or heavies, if any are separated and feed gas is spilt and sent to multiple trains for processing. In order to ensure controllability and operability of multiple trains, a thorough understanding of process dynamics and control of inlet facility is essential at an early stage of the project.

Dynamic simulation utilizing rigorous mathematical models has become an influential tool in the process design, design validation, control system verification, startup support and troubleshooting. Many recent developments in the simulation software and technology has led to the development of large scale dynamic models and their use in the entire project life.

The LNG plant in the study was receiving feed from offshore source through a long high pressure pipeline. The slug catcher pressure could operate by a differential of 50 barg during various conditions from normal to a packed line operation. The LNG plant throughput was controlled by a combination of pressure and flow controllers. The pressure control station in the feed line should maintain a steady upstream pressure for the feed flow control valve in the LNG trains.

The objective of this study was to determine the stability and controllability of the inlet feed gas system around the pressure control and feed flow during upset scenarios. One of the main objective was to identify any interaction between two pressure control valves in the inlet line. Another objective was to identify if the volume between these pressure control valves was sufficient for a smooth operation. These controllers and control valves should be able to operate smoothly under multiple transient operations. Dynamic simulation was identified as a key tool to evaluate the control configuration, valve adequacy and piping system design.

Various normal and upset transient scenarios were identified for evaluation of the design. Some of the scenarios evaluated were

- Loss of LNG Train(s)
- Loss of Domestic Gas plant
- Turndown of LNG Train
- Set point change of controllers
- Control Mode Change
- Startup of LNG Train to Turndown rate

The identified improvements in design during the FEED phase of the project were implemented during the detailed engineering phase.

(180C) Operational and safety aspects of LNG plants through dynamic simulations: disturbance scenarios, storage, and flare networks

Dimitrios Georgis[2], Matthew Okasinski[1], Pieter Schmal[3], Amit Goda[3] and Praveen Lawrence[4]
[1]Air Products and Chemicals, Allentown PA, USA.
[2]Process Systems Enterprise Inc., Houston TX, USA.
[3]Process Systems Enterprise Inc., Cedar Knolls NJ, USA.
[4]Process Systems Enterprise Ltd., London, UK.

Abstract: Physics-based dynamic models are essential tools to analyze and understand various operational and safety aspects of LNG facilities. This is illustrated through three case studies where dynamic modeling has been used to a) analyze the operational challenges during trip and scheduled maintenance events, b) assess the occurrence of rollover phenomenon within storage tanks, and c) identify, assess, and propose safety procedures during unexpected emergency events. The results demonstrate the significant value these tools offer, allowing for more confident decisions and suggesting operational changes and improvements to LNG facilities.

1. Introduction

LNG is an attractive technology for transporting natural gas across the globe mainly due to the flexibility it offers to transport natural gas to different parts of the world. LNG production is expected to grow in the next decades in order to satisfy the increasing demand for natural gas globally (BP 2017). Therefore, new LNG facilities are expected to come online in the near future in order to support the growing need for natural gas (International Gas Union 2016). Due to the energy intensive nature and the operating conditions of the LNG process, significant efforts have been devoted to analyze and improve the operation during normal, abnormal, and safety events.

The analysis of the above operational scenarios requires the development of rigorous first-principle mathematical models coupled with accurate thermodynamics that can be used for extensive dynamic simulations under various operational scenarios. This necessity can be fulfilled through the adoption of advanced process modeling technology. gPROMS is an equation-oriented advanced process modeling platform offering significant value to various process industries. One of the main features of the gPROMS platform is the capability to perform dynamic simulations involving complex operating procedures of models of large-scale processes that are tightly integrated and have multiple recycles. Using this tool, three different applications demonstrate the value that dynamic analysis offers during various operational aspects of LNG facilities.

In the next section, the first study focuses on the dynamic behavior of a liquefaction process with parallel refrigerant compressor strings. Parallel configurations present unique operating challenges when a trip or scheduled maintenance event takes a string offline or when the same string needs to be returned to service without disrupting current operations. The second case study demonstrates how dynamic modeling can be used to accurately assess the LNG rollover in storage tanks; the rollover phenomenon occurs when rapid mixing of LNG of different conditions takes place within the storage tank, leading to large amounts of vapor being released that could lead to the over-pressurization of the tank. This can result in a severe accident as was the case in LaSpezia, Italy in 1971 (Sarsten 1972). The third case study focuses on the process safety under emergency events in LNG facilities. During these events the depressurization of process equipment and piping, namely blowdown, is a critical operation and requires consideration. This case study demonstrates how a rigorous dynamic assessment is used to optimize the blowdown operation for LNG facilities, reducing blowdown duration, and delivering a safer facility.

2. The value of using rigorous dynamic analysis in various operations of LNG facilities

a) Operational challenges in facilities with parallel refrigeration strings:

In recent years, some baseload LNG projects have chosen to use parallel refrigeration strings to improve turndown efficiency, compressor aerodynamics, and on-stream availability. However, this configuration introduces operating consequences such as how to smoothly transition between (n) string operation and (n-1) string operation without causing a sympathetic trip of the surviving refrigeration string(s). In addition, the transition between (n-1) strings to (n) string operation also needs careful consideration during the detailed design phase of the project.

This section of the paper will first discuss the issue of a trip one of two or more parallel strings. Losing one of two strings represents a 50% loss of power into the process. Because LNG production is proportional to the power consumed by the refrigeration system, the new steady-state production after the transition will be approximately 50% of design production. As (n) increases, there is an increased equipment cost due to duplicating the refrigeration string (n-1) times but the production lost by the trip of one string decreases by approximately maximum production / (n). As two parallel refrigeration strings is the most common configuration and causes the largest change in available refrigeration, the example will focus on a liquefaction process with two parallel mixed refrigerant strings. However, the learnings also apply to (n) > 2.

Shown schematically in Figure 1 is a propane – mixed refrigerant (AP-C3MR™) process with two parallel mixed refrigerant (MR) and propane (C3) compressor strings. For simplicity, the two propane compressor strings are shown as a single box. Dry, clean natural gas is compressed to high pressure and then cooled against propane prior to entering the Main Cryogenic Heat Exchanger (MCHE). Within the MCHE, the feed gas is further cooled, liquefied, and subcooled against a mixed refrigerant containing nitrogen and light hydrocarbons. Upon exiting the MCHE, the LNG stream is reduced to nearly atmospheric pressure across, for example, a hydraulic turbine, valve or some other similar device. The mixed refrigerant (MR), carrying the heat removed from the tube circuits, exits the MCHE shell, is compressed to high pressure, and cooled against propane. After phase separation, the mixed refrigerant liquid (MRL) and vapor (MRV) are further cooled and liquefied within the MCHE tubes.

Finally, the MRV and MRL are reduced in pressure across the warm and cold Joule Thompson valves (WJT and CJT respectively) to provide refrigeration for the MCHE.

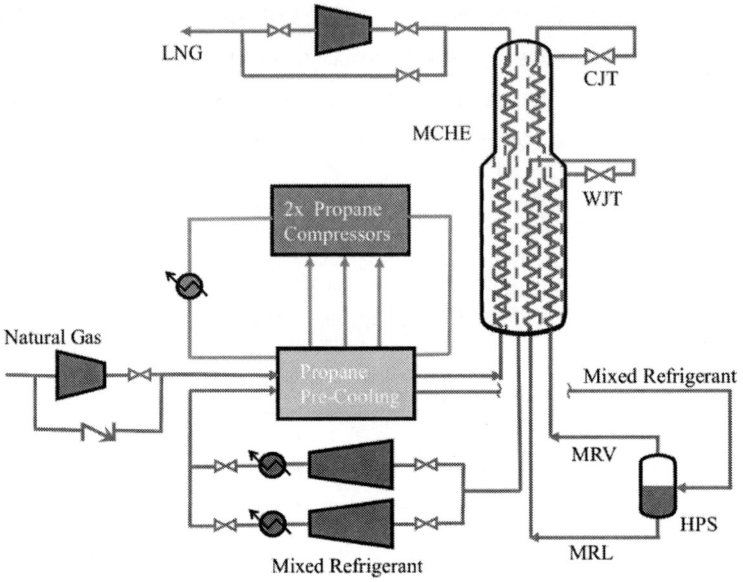

Figure 1: AP-C3MR™ Schematic

In the event of an MR string trip, there are several triggers that will cause a sympathetic trip of the surviving string. These triggers include:

1. warm LNG temperature due to refrigeration imbalance;
2. LPMR compressor suction temperature gets too cold; and/or
3. the surviving compressor string attempts to compress the entire MR flow, becomes overpowered, and the turbine slows to its low-speed trip point.

The goal of the dynamic simulation is to demonstrate a transition of the liquefaction process from full production to reduced production without reaching any of these triggers. As mentioned above, the sudden loss of one of two refrigeration strings means that the process can only produce approximately 50% of design production. So upon the trip signal, LNG flow must be immediately reduced by partially closing the LNG valve to a predetermined opening. This prevents warm LNG exiting the MCHE, which would in turn, trip the liquefaction unit (trigger #1 above).

On the refrigeration side, the MR flowrate must also be reduced in approximate proportion to the LNG flowrate (Figure 2) to prevent a low shellside outlet temperature (trigger #2). So simultaneously with the feed reduction, both the warm and cold JT valves partially close to a predetermined opening. Once the refrigeration and heat load begin to rebalance, the shellside outlet temperature stabilizes, as seen in Figure 3. These predetermined settings are calculated based on multiple dynamic simulations and then fine-tuned in the field, based on plant performance.

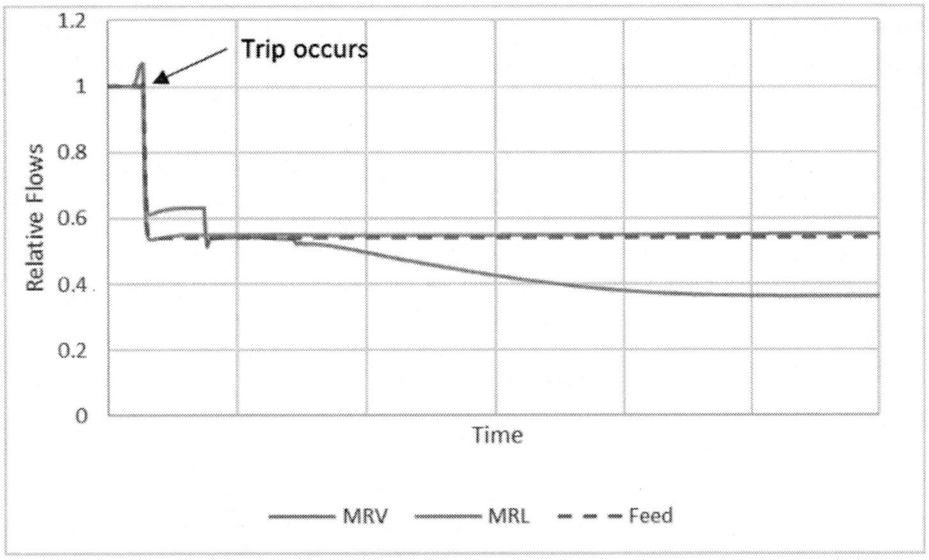

Figure 2: MCHE Flowrates

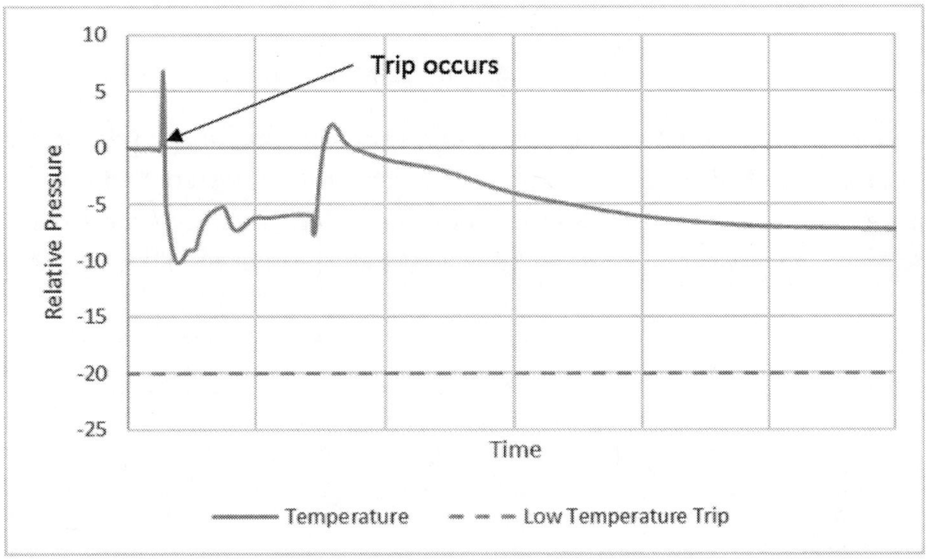

Figure 3: MCHE Shellside Outlet Temperature

However, simply balancing heat load and refrigeration is not enough to ensure a smooth transition to a reduced production. When a compressor string is suddenly isolated, the flow that was going into (n) strings can now only flow through (n-1) strings. This pushes the operating point of the surviving strings' LP compressor towards stonewall (Figure 4), increasing the power demand from its turbine. Once the turbine reaches its maximum available power, it will slow down. A gas turbine has limited speed range and could trip on low speed (trigger #3).

Through use of dynamic simulation, a methodology was developed to transition from (n) to (n-1) strings without a sympathetic trip. The goal is to keep the compressor within its normal operating range between surge and stonewall (*Figure 4*). This is accomplished by quickly *reducing* the flow into the surviving string's first stage compressor suction by throttling the inlet valve. As with the WJT and CJT valves, dynamic simulations are used to determine the speed and degree of throttling to keep the surviving string operating within its normal range.

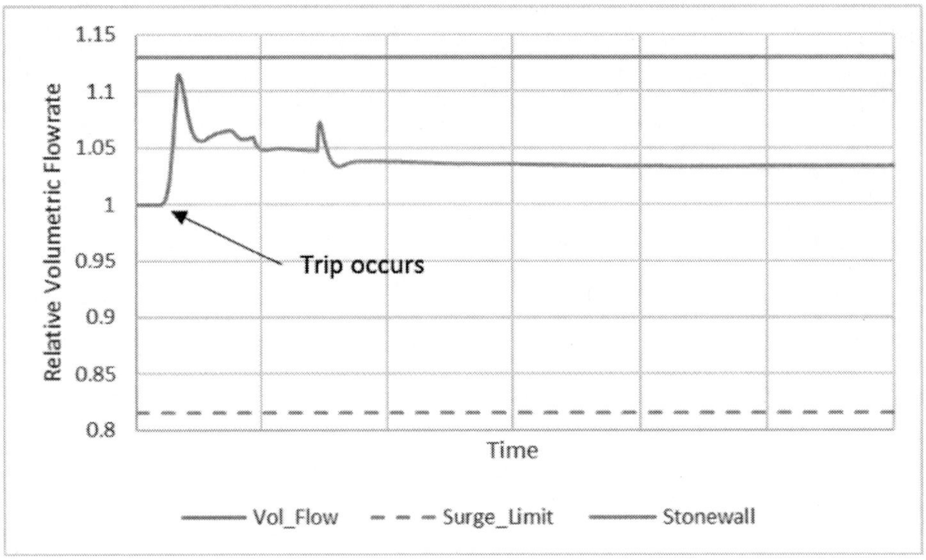

Figure 4: Surviving String LPMR Compressor

The consequence of suction throttling *temporarily* raises the pressure in the shellside MCHE, as seen in Figure 5. This is one of the largest capacitances within the MR system and operates well below its design pressure. So it can store refrigerant until the surviving compressor string has had time to compress the excess refrigerant in the MCHE shell and store the unused MRL in the MR Separator.

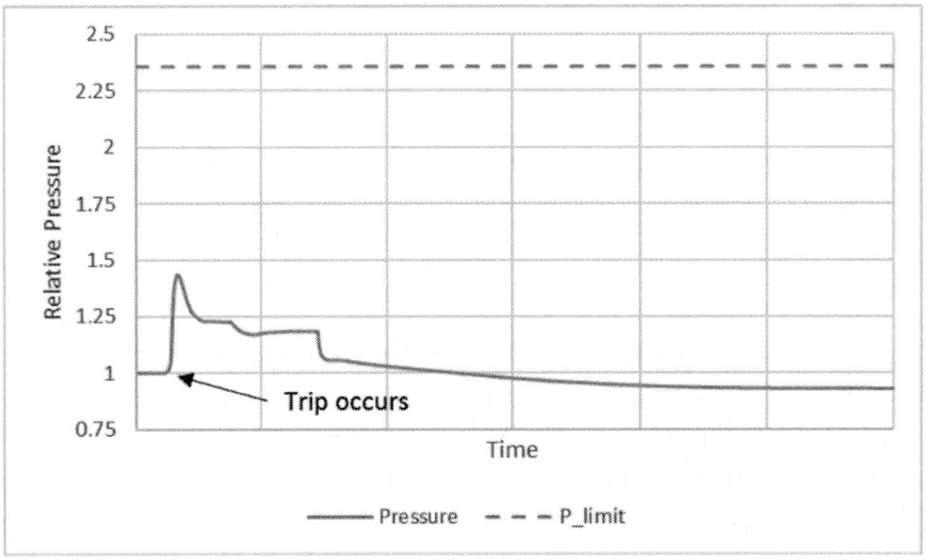

Figure 5: MCHE Shellside Outlet Pressure

This response strategy shows the value of dynamic simulation. One's first thought is that after losing 50% of the MR compression capacity, the best strategy is to use the remaining compressor to compress as much flow as possible. However, this approach is expected to cause a sympathetic trip. Rather, the correct action is to keep the flow to the compressor within an acceptable operating range, preventing overloading the gas turbine, and then slowly moving the extra MR from the MCHE and storing it in the MR separator.

Throughout this disturbance, the LNG outlet temperature (Figure 6) stayed within a few degrees Centigrade of its pre-trip value. Note that the LNG stayed colder throughout the transition to a lower production. With one compressor string offline, there is a reduced need for fuel gas, so the pre-determined positions of the refrigeration valves were chosen to yield a colder LNG.

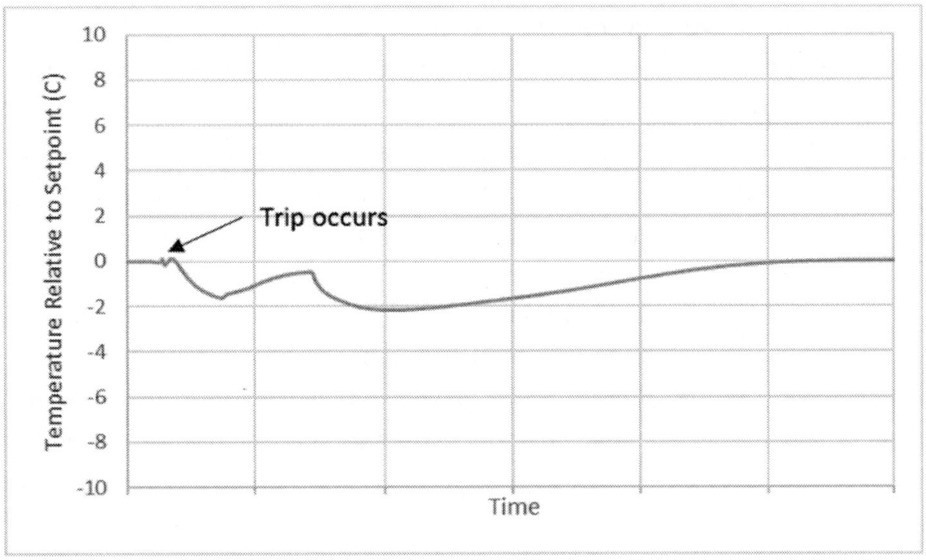

Figure 6: LNG Temperature

Parallel refrigeration strings also present operational challenges when restarting an offline string while the other string(s) are already online and the process is operating at reduced production. In particular, the offline string is on full recycle from the aftercooler outlet, so its inlet temperature is near ambient. The online LPMR suction drum is at approximately -35 °C. The warmer inlet temperature will cause the outlet pressure of the offline string to be lower than the online string, so the inlet temperatures must be matched to bring it online. Various options for cooling the offline suction are discussed in Okasinski 2016.

These types of dynamic simulations were performed for a currently operating baseload plant in addition to several LNG plants currently under construction, all containing parallel compressor strings. Initial feedback shows that the model accurately predicts plant behavior.

b) <u>Accurate assessment of rollover phenomena in LNG storage tanks:</u>

The cryogenic nature of LNG introduces additional challenges during transportation and storage. LNG is typically stored at very low temperatures (e.g., -163 °C) in well-insulated storage tanks. However, heat entering (heat ingress) from the environment into the storage tank causes a slow evaporation (boil-off) of the more volatile components in LNG (e.g., nitrogen, methane). This causes a change in the composition in the tank, a process called weathering. The weathering process leads to a higher density LNG, and due to boil-off at the surface, the temperature of the LNG varies. The density and temperature difference of the LNG in the LNG tank causes circulation due to natural convection in the tank.

When newly-produced LNG is introduced into the storage tank, fill-induced stratification may result. This phenomenon is attributed to the different conditions (density and temperature) of the already existing LNG in the tank and the LNG being added. If the stratification is significant, the LNG at the lower layer (heavy LNG) can become superheated mainly due to the heat accumulated from heat ingresses within the lower part of the storage tank since the stratified layer above does not allow the heat to escape. The composition of the LNG in the stratified layer also varies due to continuous boil-off of the light components in LNG in the upper layer and heat and mass transfer between the stratified LNG layers. When the densities of the stratified LNG layers become equal, the interface between the stratified LNG layers becomes unstable and rapid mixing happens with the lower layer releasing all of its accumulated heat. This phenomenon is called rollover. Rollover can result in significant vapor release which can result in significant loss of valuable fuel and/or over-pressurization of the storage tank leading to a potential structural damage or failure. Therefore, it is evident that assessing the time the rollover phenomenon occurs is crucial.

The mathematical model developed by Heestand et al. (J. Heestand 1983) was used as a starting point to model the dynamic behavior and assess the occurrence of the rollover phenomenon within the LNG storage tank. The mathematical model developed by Heestand et al. has been validated against results from the LaSpezia rollover incident in Italy in 1971. In this case study, we improved the fidelity of the above model by relaxing some of the assumptions and consider physical phenomena that were previously ignored. For example, we considered non-ideal behavior of the mixture for a more accurate prediction of the thermodynamic vapor-liquid equilibrium; we relaxed the assumption of constant pressure across each layer and accounted for the heat transferred due to mass transfer. The modified model was validated against the results of the LaSpezia rollover incident to assess how accurately it predicts the onset of the rollover.

In Figure 7 the transient responses of the densities of each layer reveal that there is a continuous mass transfer due to the different densities until about 31 hours at which time the two densities become equal, a necessary condition for the onset of the rollover phenomenon. This model is in good agreement with both the results from LaSpezia rollover incident and the model developed by Heestand et al. in terms of predicting the rollover time. In Figure 8 the transient responses of the temperature for each layer demonstrate clearly that, at the time of the rollover, there is a temperature difference of more than 3 Kelvin. The larger this temperature difference, the more severe the rollover is expected to be (higher boil-off rates). In Figure 9 the boil-off gas rate predicted using the gPROMS model is plotted against time and compared with the data available for the LaSpezia incident.

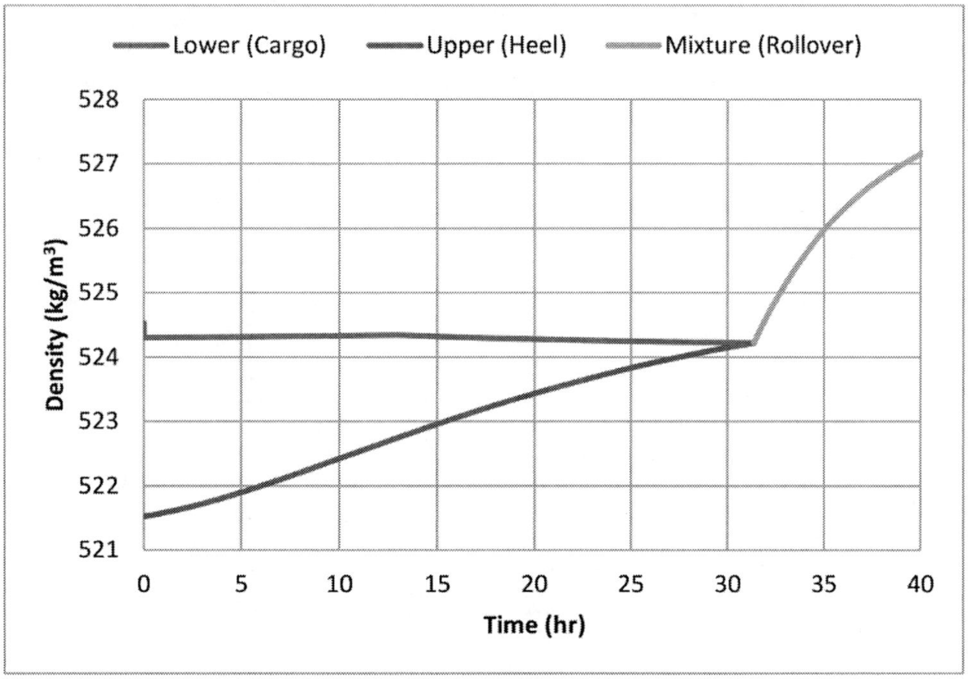

Figure 7: Densities of each layer in the LNG storage tank

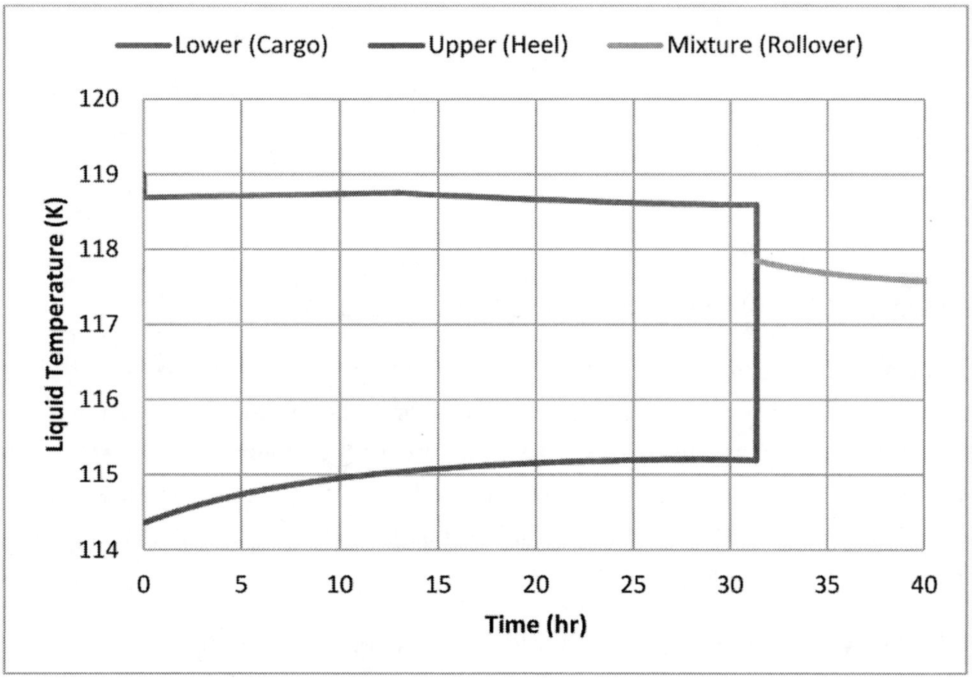

Figure 8: Temperature of each layer in the LNG storage tank

This case study demonstrated how dynamic analysis can be used to either analyze the current status of the LNG storage tank or even explore alternative strategies to prevent rollover phenomena. In addition to this, developing lumped parameter models to describe this physical phenomenon in LNG storage tanks is mainly adequate for predicting the time at which the rollover phenomenon happens. Although the above models can also be used to predict the behavior after the rollover, they may not be very accurate since different physical phenomena (which are not captured in the lumped parameter models) dominate the behavior of the LNG within the storage tank at these conditions.

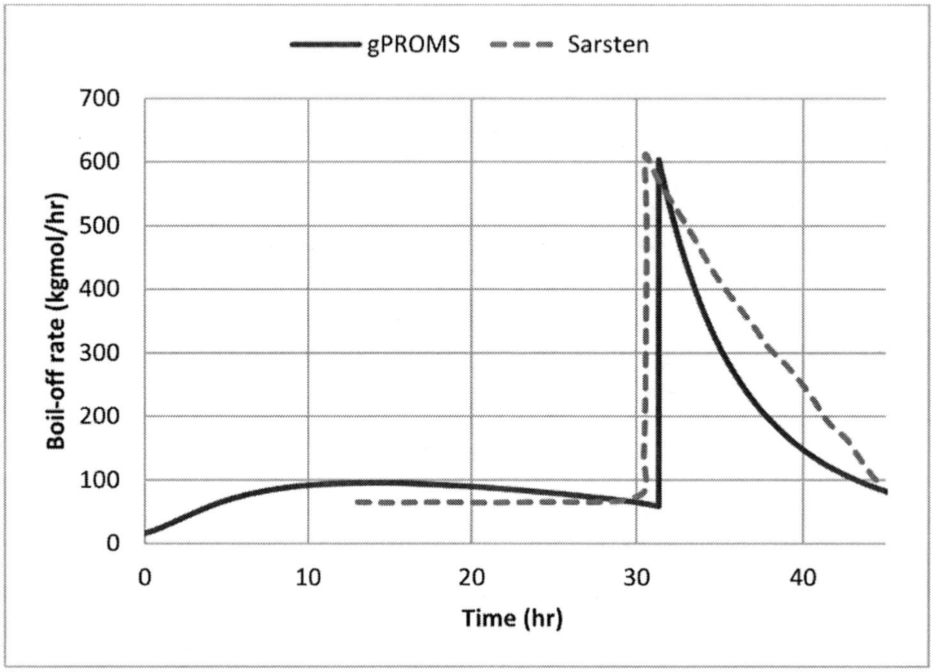

Figure 9: Boil-off rate prediction between gPROMS and Sarsten

Therefore, a different modeling approach should be followed if one is interested in analyzing the behavior following the rollover.

c) <u>Improve operation under two different emergency events:</u>

A blowdown (depressurization of process equipment and piping) is a safety critical operation in a LNG plant and is required to ensure safe disposal of hydrocarbon inventory during a plant emergency. During a plant shut-down or emergency situation, the entire plant is segregated into a number of blowdown segments and their inventory depressurized at a fast rate to permit reduction of the vessel stress to a level at which stress rupture is not of immediate concern (to reduce the consequences from a vessel leak or failure during a fire).

For a safe blowdown operation, the hydrocarbon inventory should be safely disposed of via the coupled flare system; consequently, the flare system design must be assessed for the worst credible blowdown scenarios to check for the following design constraints:

- Depressurization time of blowdown segments;
- Flare design pressure;

- Minimum design temperature constraints both within process segments and flare piping;
- Flare tip design flowrate;
- Slugging, acoustic, and flow induced vibration risks; and
- Noise and radiation constraints.

The typical worst credible blowdown scenarios in a LNG plant are the following:

- Blowdown of an emergency depressurization (EDP) zone when the segments within the zone are exposed to fire; and
- Total power loss.

A blowdown operation is a fast transient operation by nature, and to add to the complexity to the process dynamics, it is not uncommon for multiple blowdown valves (BDVs) in a LNG plant to have staggered opening and staged opening of the flare tips.

In spite of this complex dynamic behavior, traditionally flare networks are designed using steady state modeling tools leading to a design that does not meet the requirements. Increasingly it is recognized that dynamic analysis is important in flare system design. Here we consider two case studies from recent major capital projects; these demonstrate the benefits of a **fully coupled dynamic flare and blowdown model-based approach**. Note that in a fully coupled dynamic flare and blowdown model, the models of all the process blowdown segments are connected to the model of the flare network and solved simultaneously.

Case 1: Simultaneous depressurization of multiple BDVs in an EDP zone (fire)

In this study: the BDVs in this plant were sized to depressurize the blowdown segment as per the typical American Petroleum Institute (API) guidelines by carrying out standalone blowdown calculations i.e., with an assumed flare back pressure of 1 bar. However, as per the design philosophy of the plant, in the event of fire, all the blowdown segments within an EDP zone must be depressurized simultaneously.

For an accurate assessment of the system behavior, a fully coupled dynamic flare and blowdown calculation was carried out.

- The calculation proved that the simultaneous depressurization of all the BDVs within the EDP zone causes a backpressure large enough to affect the depressurization rate of one of the blowdown segments – this is shown in Figure 10, which compares the results from a conventional stand-alone calculation to that from the fully coupled dynamic analysis.
- The fully coupled dynamic flare and blowdown calculation (see Figure 10) clearly shows that the large back pressure caused the process blowdown segment design violations (segment depressurized to 50% of design pressure in approximately 19 minutes compared to 15 minutes using standalone calculations) but relying on the standalone calculations would led to the incorrect conclusion that the depressurization rate of the segment is adequate. In this case, to

rectify this design issue, further checks were made to verify if the segment depressurization rate is fast enough to reduce the internal stress within the process vessels to acceptable levels in the event of fire, and suitable mitigation strategies were put in place.

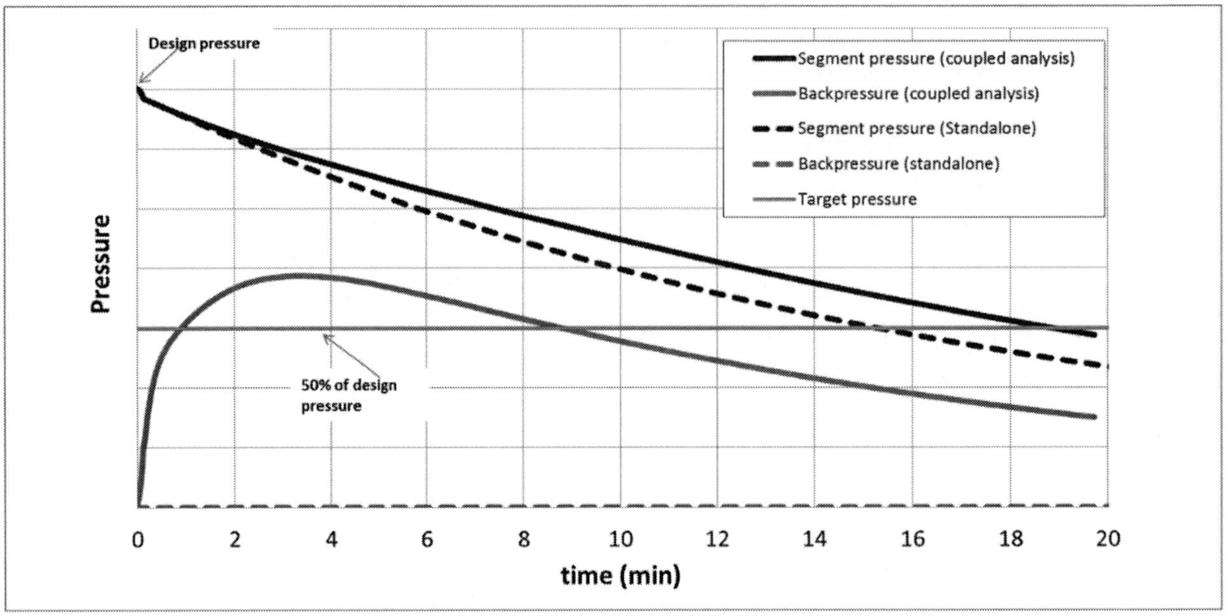

Figure 10: Comparison of blowdown profile for one blowdown segment within an EDP zone. Calculation carried out in standalone mode, and fully coupled flare and blowdown modeling approach

Case 2: Depressurizing multiple EDP zones in the event of total power loss in a LNG plant

During plant operation, if the main power generation equipment at the LNG plant were to fail and the emergency generators also fail to sustain the electrical loads, then there would be a limited operation window (typically a couple of hours) to utilize the Uninterruptible Power Supply (UPS) to safely shutdown the entire plant. Failing to shutdown the entire plant would cause all the BDVs in the plant to de-energize simultaneously and overload the flare network.

Ideally, in the event of loss of power in the plant, the objective is to restart the emergency generators/main power as soon as possible, but if this is not possible, all the hydrocarbon inventory in the EDP zones in the plant must be depressurized before the UPS is depleted. As a simultaneous depressurization of all the EDP zones in the plant can lead to flare overloading, a staggered blowdown philosophy is typically adopted in LNG plants and this approach can take hours to fully depressurize the plant.

The conventional design approach is to carry out data processing on the stand-alone blowdown profiles of all the blowdown segments in the plant i.e., stitching together individual blowdown profiles to define a conservative staggering recipe that ensures that the peak flowrate entering the flare network is always below the flare design flowrate. While this approach may lead to safe depressurization of the system, it typically results in an excessively large flare system, and so for a given flare system size, grossly underutilizes its capacity and significantly reduces the time available for the operator to take corrective action to restore power in the plant.

A fully coupled dynamic flare and blowdown model is required for an optimal design of the staggering recipe that minimizes the time required to depressurize the plant without overloading the flare network. This approach is essential to capture both the process dynamics within the blowdown segments and the flare dynamics/packing effect in the flare network to provide an accurate assessment of the staggering recipes.

Figure 11 shows the comparison of the staggering recipe profile for the conventional approach using a plurality of standalone assessments (i.e., stitching together individual blowdown profiles) and the more accurate fully coupled dynamic flare and blowdown approach. A design based on the conventional solution approach would have led to the conclusion that it would take approximately 1.5 hours to depressurize the plant, but the fully coupled dynamic flare and blowdown analysis clearly demonstrates that the plant flare network has sufficient capacity to depressurize in approximately 45 minutes, leaving additional time for the operator to take any corrective action in the plant. Note that the new staggering recipe shows that there is even more capacity in the flare network, and if required, the staggering recipe can be altered to reduce the depressurization time even further.

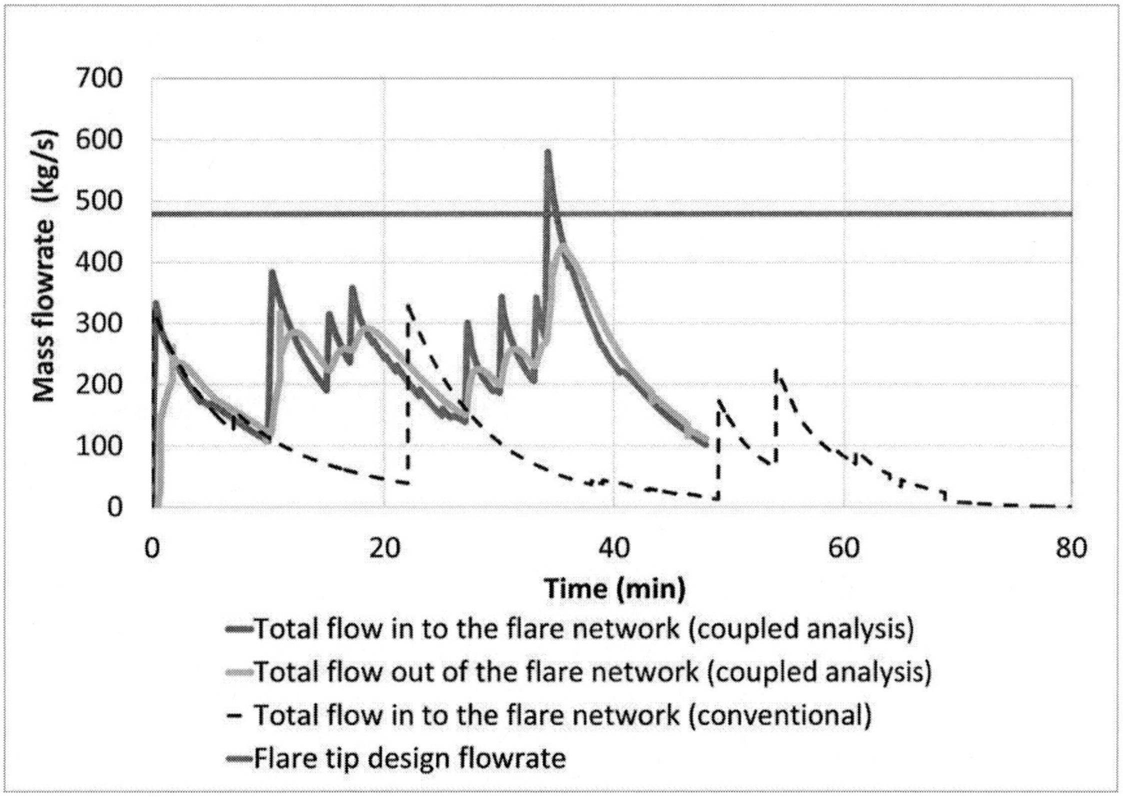

Figure 11: Comparison of flare flowrate profile for total power failure scenario using the conventional approach, and fully coupled flare and blowdown modeling approach.

3. Conclusions

This paper demonstrates the value generated by using dynamic simulations of mathematical models based on first-principles to analyze operational and safety aspects of LNG facilities. As discussed in the paper, although the parallel refrigerant compressors provide operating flexibility, they present unique challenges that need to be understood and resolved during the engineering phase. A dynamic model of the liquefaction process provides both insight into disturbance response and can determine preliminary control actions to transition the process between stable operating points. The second case study illustrated the contribution of dynamic modeling in accurately predicting the occurrence of rollover phenomena and the opportunities for further analysis on preventing it. The last case study revealed the importance and the benefit of dynamically analyzing the behavior and improving the operation of integrated systems (process and flare) under different emergency events.

References:

BP. "BP Energy Outlook 2017 Edition." 2017.

International Gas Union. "2016 World LNG Report." 2016.

J. Heestand, C.W. Shipman, J.W. Meader. "A Predictive Model for Rollover in Stratified LNG." *AIChE Journal* 29, no. 2 (1983): 199-207.

Okasinski, et al. "Recommendations for Restarting a Parallel Compressor String Based on Dynamic Simulation." *Proceedings of LNG 18 Conference.* Perth, 2016.

Sarsten, J.S. "LNG Stratification and Rollover." *Pipeline and Gas Journal* 199 (1972): 37-42.

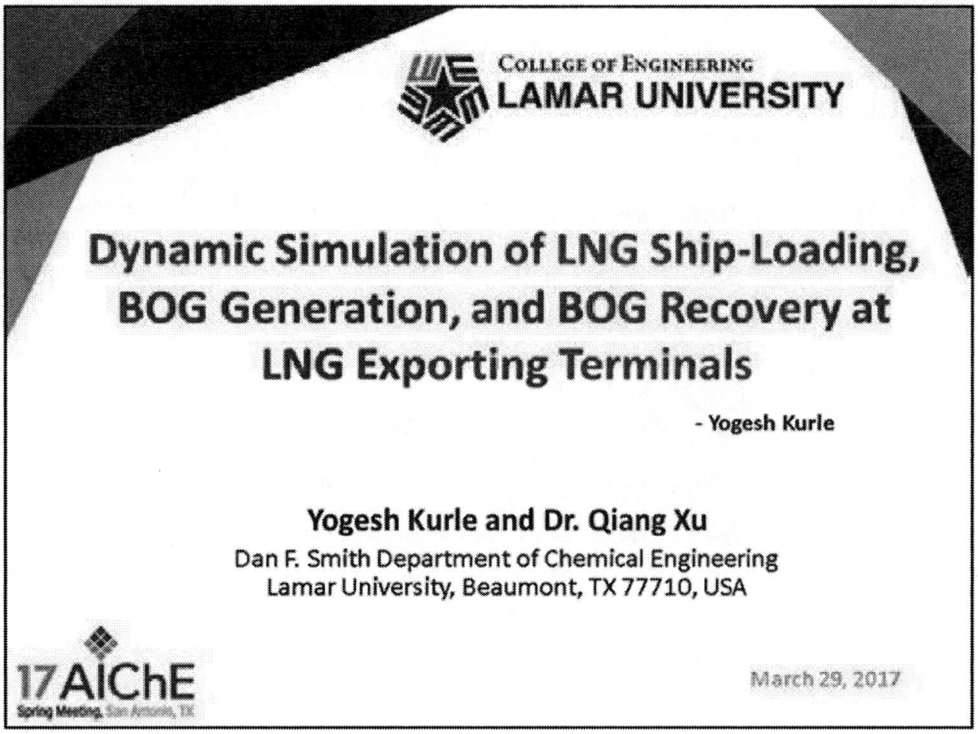

- ❖ In this work, the LNG vessel loading process is dynamically simulated to obtain Jetty BOG generation profiles.

Presentation Outline

- Introduction
- Natural Gas Supply Chain
- Jetty Boil-Off Gas
- Process Simulation and Controller Setup
- JBOG profile with considerations of Holding Mode Heat Leak
- JBOG profile for different initial-temperature of ship-tanks
- JBOG profile for different JBOG-compressor capacity
- JBOG profile for different maximum-allowable cooling-rate
- Conclusion

Introduction

- Natural Gas contains methane as major component
- Natural Gas is gaseous fossil fuel
- It is a colorless, odorless, non-corrosive, non-toxic and cleanest fossil fuel
- Emits less than half CO_2 as compared to coal emission
- Heating value = 37 MJ/m^3, NG-fired co-generation conversion efficiency 57% 57%
- Abundant reserves
- Being clean, safe and reliable, NG is becoming major source of energy
- Used for heating, cooking, power generation
- NG is world's fastest growing energy economy

17 AIChE Spring Meeting, San Antonio, TX

3

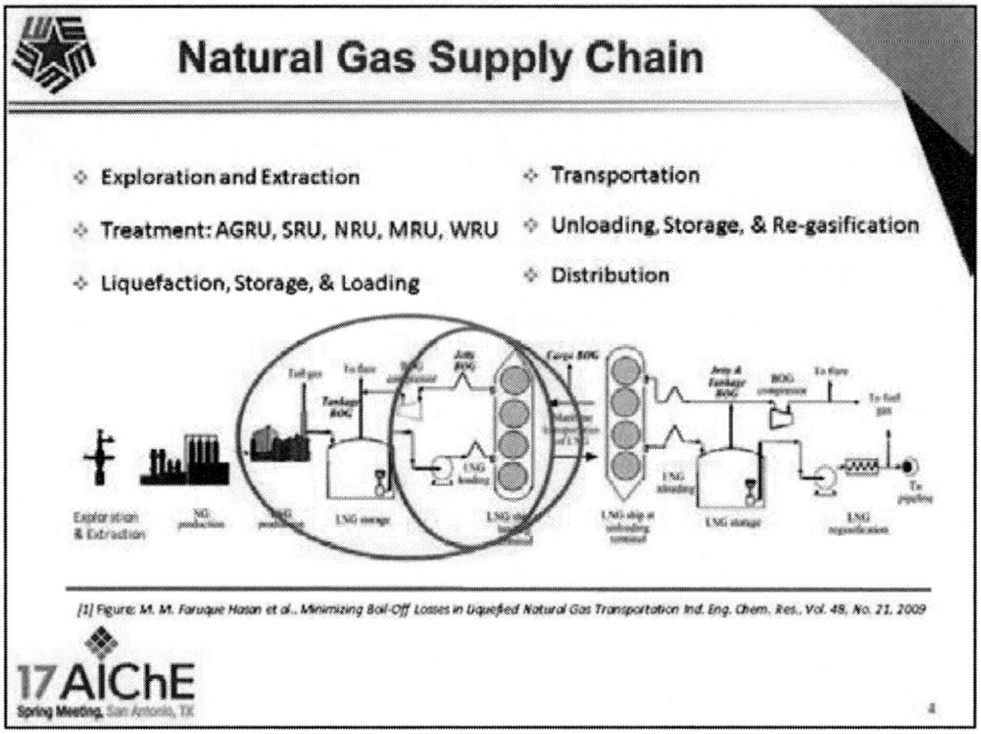

- ❖ In the Natural Gas supply chain, our focus of study is on LNG exporting terminal, more specifically on LNG loading section. LNG loading into LNG ship generated Jetty BOG (JBOG).
- ❖ The rate of JBOG generation during LNG loading changes with loading time. Thus, JBOG generation needs to be well studied so that it can be handled properly to avoid potential process upsets, flaring and wastage of material. Because of the dynamic nature of the LNG loading process, detailed dynamic simulations need to be employed to understand the insight of the process.

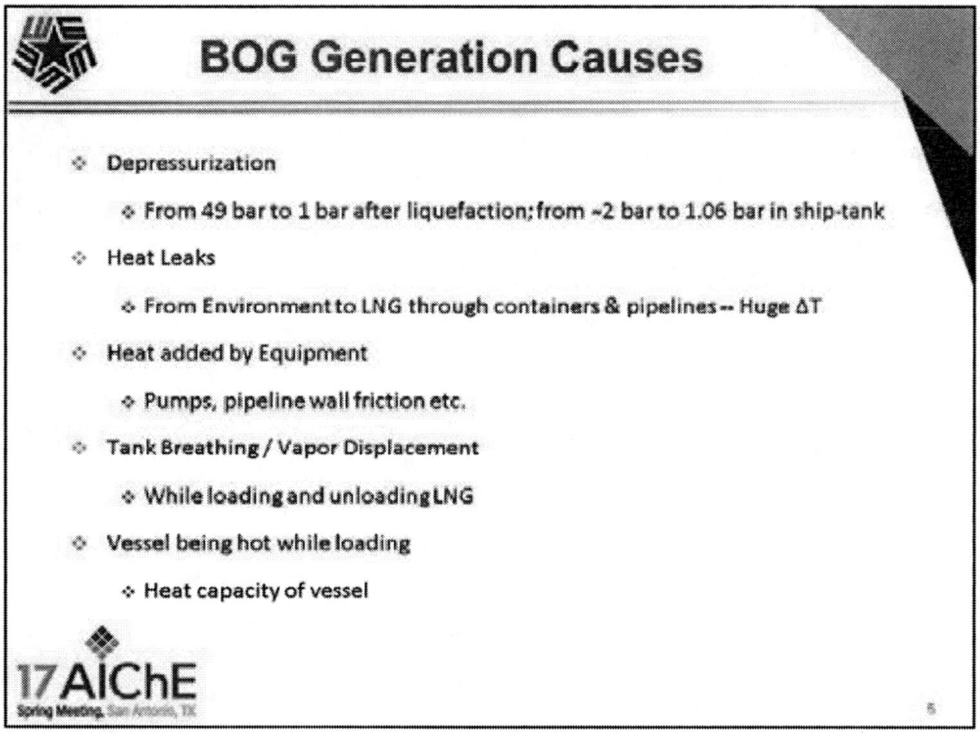

- BOG generation is caused by several factors: (1) depressurization of LNG (flashing); (2) environmental heat leaks through containers and pipelines; (3) heat added by equipment like pumps; (4) tank breathing or vapor displacement; and (5) LNG carrying vessels being relatively hot while loading LNG.
- Heat leak from environment into LNG occurs continuously since there is always difference in temperature of ambient and temperature of LNG. The heat leak from hotter tank into LNG is due to heat content of the metal of the tank, which vanishes once thermal equilibrium state is achieved between the metal and LNG.

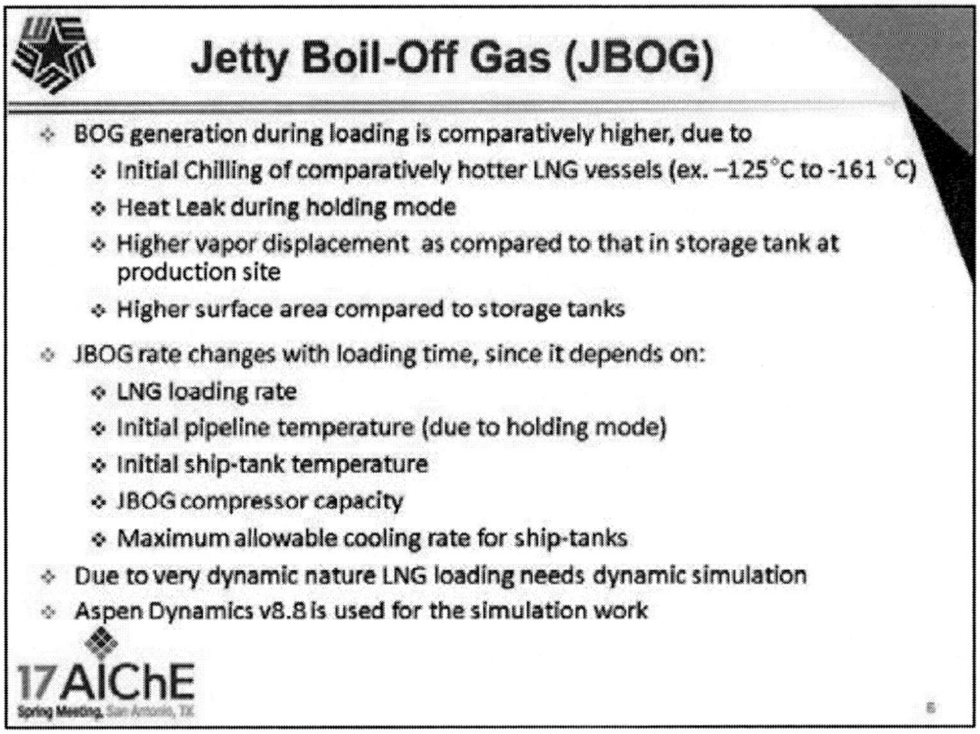

- ❖ BOG generation is relatively more at exporting terminals
- ❖ BOG generation during LNG ship-loading is very dynamic due the factors mentioned. Therefore it is important to study the effects of these factors on JBOG.

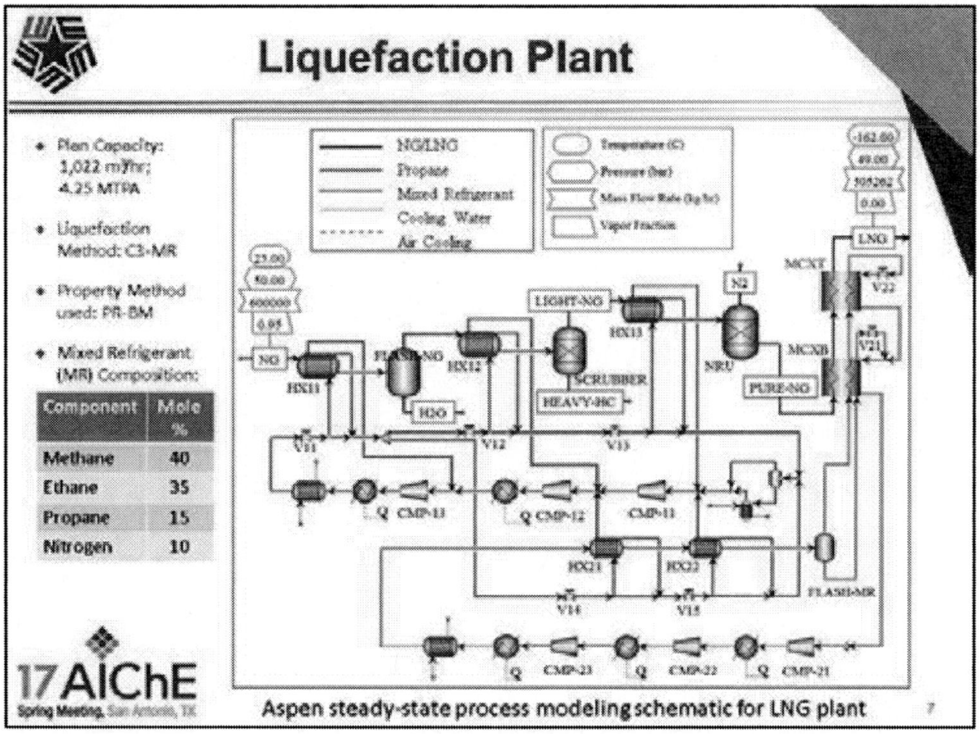

❖ A typical Propane-and-Mixed-Refrigerant (C3-MR) process by Air Products and Chemicals Inc. (APCI) was used for liquefaction of natural gas. Above process flow shows Natural gas liquefaction section.

❖ The steady state process was simulated using Aspen Plus v8.8, and exported to Aspen Plus Dynamics v8.8 to study the dynamic behavior of LNG loading facility.

❖ Peng Robinson cubic equation of state with the Boston-Mathias alpha function (PR-BM) property method was used for the process simulation.

❖ The process parameters for the liquefaction section were taken from article by Ravavarapu et al.(Ravavarapu et al., 1996)

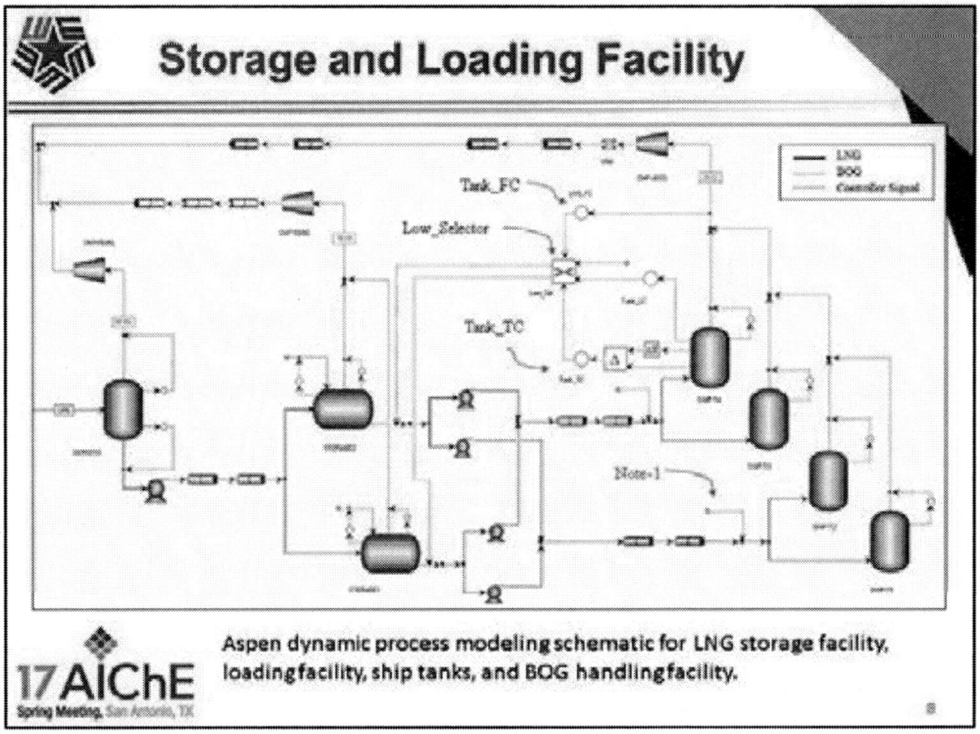

- Above process flow shows LNG storage facilities, and loading facilities.
- Three main BOG generation locations are identified at LNG exporting terminals: (1) Flash Tank after the main cryogenic heat exchanger (MCHE) – FBOG is generated here, (2) Storage-Tanks – TBOG is generated here, and (3) Jetty – JBOG is generated at this location.
- In order to satisfy JBOG compressor capacity constraint and ship-tank cooling rate constraint, two controllers namely LNG_FC and Tank_TC are set up. When ship-tank level reaches 80 vol%, the loading rate is ramped down by the script (Task) written in the simulation. The Low-Sel block in Figure selects lowest of these values – output of LNG-FC, output of Tank_TC, and value chosen by the Task. This way each constraint is satisfied with just one manipulated variable – LNG loading rate. For a particular instance, whichever constraint has dominating effect on loading rate will require lowest loading rate.
- Note 1 denotes where holding mode heat leak was added in the simulation

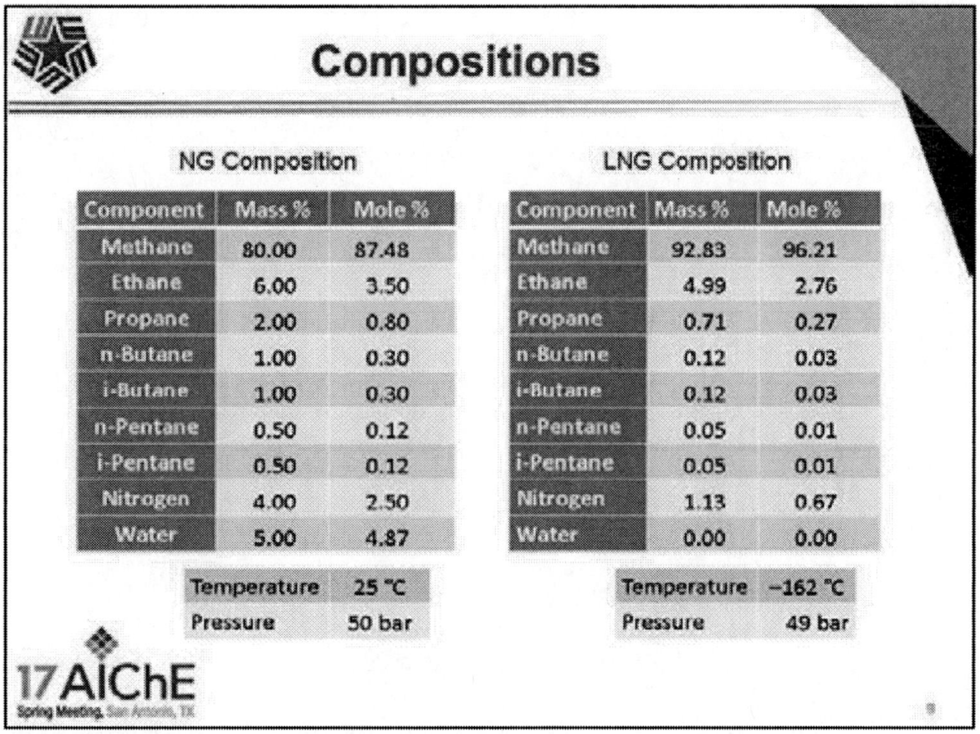

Compositions

NG Composition

Component	Mass %	Mole %
Methane	80.00	87.48
Ethane	6.00	3.50
Propane	2.00	0.80
n-Butane	1.00	0.30
i-Butane	1.00	0.30
n-Pentane	0.50	0.12
i-Pentane	0.50	0.12
Nitrogen	4.00	2.50
Water	5.00	4.87

Temperature	25 °C
Pressure	50 bar

LNG Composition

Component	Mass %	Mole %
Methane	92.83	96.21
Ethane	4.99	2.76
Propane	0.71	0.27
n-Butane	0.12	0.03
i-Butane	0.12	0.03
n-Pentane	0.05	0.01
i-Pentane	0.05	0.01
Nitrogen	1.13	0.67
Water	0.00	0.00

Temperature	−162 °C
Pressure	49 bar

- ❖ These are the natural gas (feed) composition, and LNG (product) composition

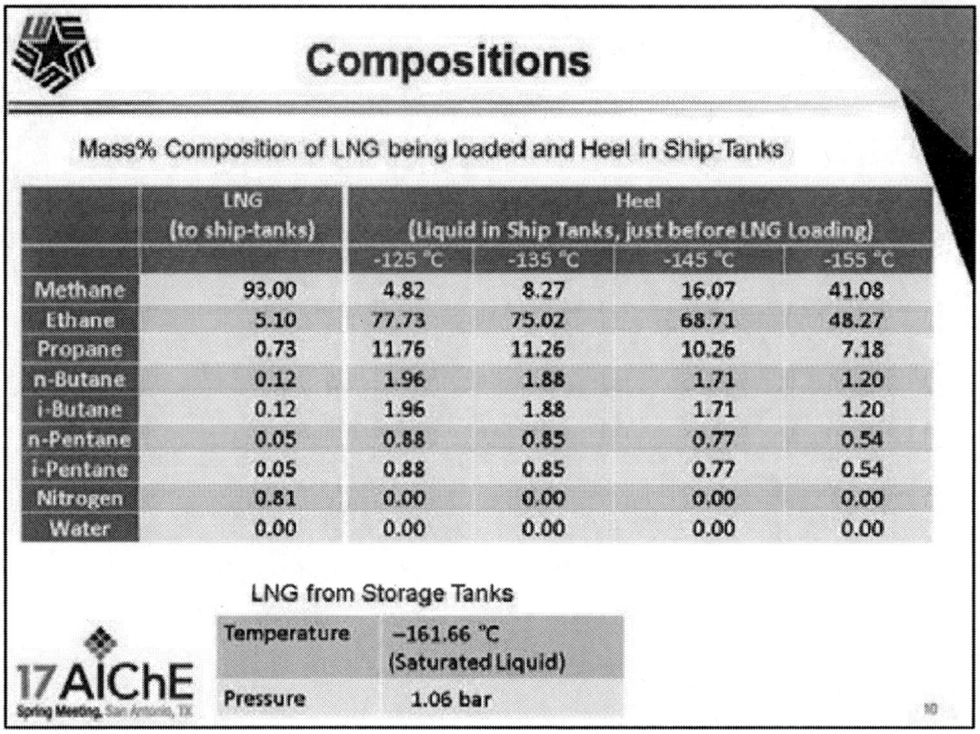

	LNG (to ship-tanks)	Heel (Liquid in Ship Tanks, just before LNG Loading)			
		-125 °C	-135 °C	-145 °C	-155 °C
Methane	93.00	4.82	8.27	16.07	41.08
Ethane	5.10	77.73	75.02	68.71	48.27
Propane	0.73	11.76	11.26	10.26	7.18
n-Butane	0.12	1.96	1.88	1.71	1.20
i-Butane	0.12	1.96	1.88	1.71	1.20
n-Pentane	0.05	0.88	0.85	0.77	0.54
i-Pentane	0.05	0.88	0.85	0.77	0.54
Nitrogen	0.81	0.00	0.00	0.00	0.00
Water	0.00	0.00	0.00	0.00	0.00

Temperature	−161.66 °C (Saturated Liquid)
Pressure	1.06 bar

- ❖ Composition of LNG being loaded to ship-tanks is shown.
- ❖ Also the hotter ship-tanks will have different liquid compositions based on temperatures before LNG loading. In each case 1 vol% heel is assumed to be remaining before LNG loading.

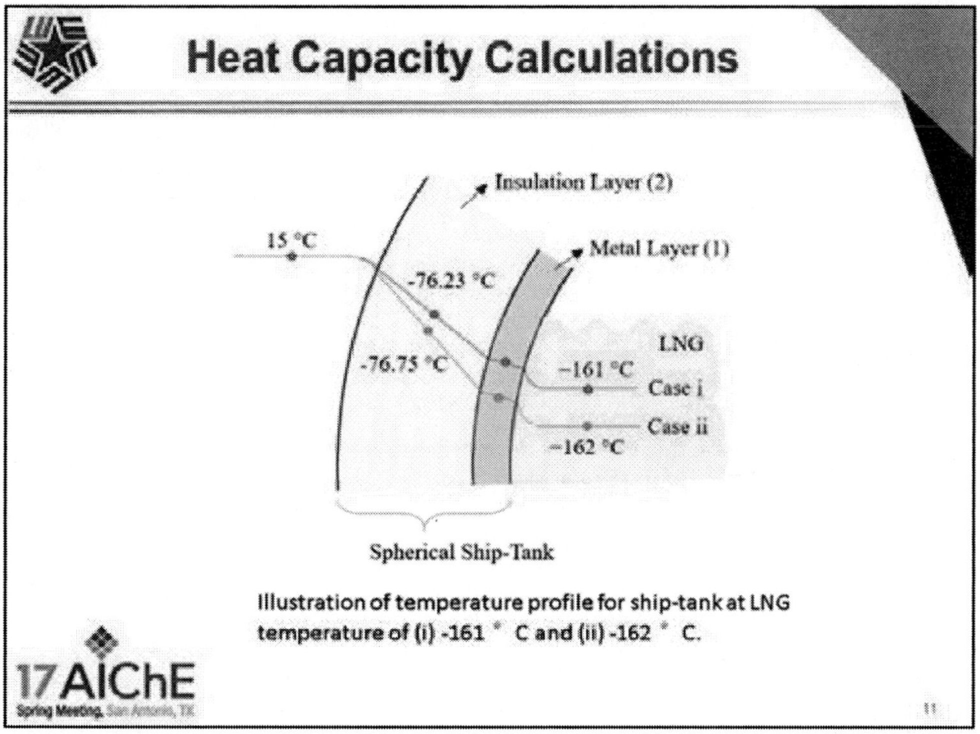

- ❖ One-degree change in LNG temperature causes almost one-degree change in inner metal layer. However, the corresponding temperature change for the outer insulation layer is significantly less than one degree. Therefore, mere addition of heat capacities of two layers would not indicate the amount of heat to be removed from the ship-tank to cool it down by 1 °C.
- ❖ Figure illustrates temperature profile of spherical wall of ship-tank for two different cases. Case i and case ii corresponds to LNG temperature of -161 °C and -162 °C respectively. When there is 1-degree change in LNG temperature, the corresponding change in average temperature of insulation layer is about 0.52 °C.

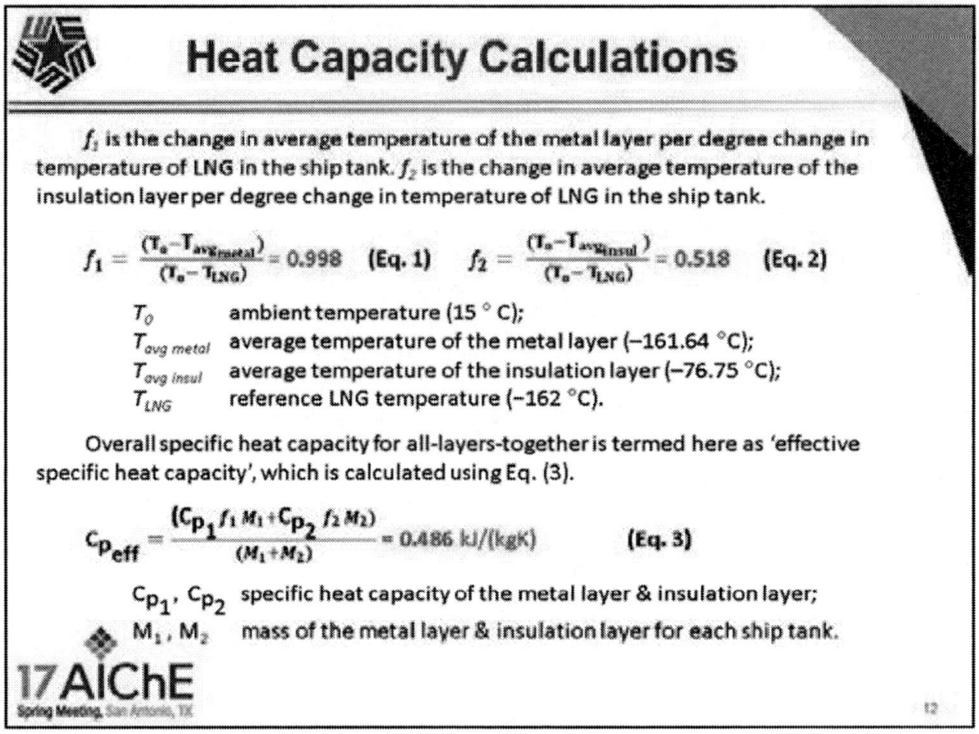

❖ Parameter f1 and f2 are defined in Eqs. (1) and (2) respectively to indicate the change in average temperature of a layer with respect to the change in temperature of LNG inside the ship tank. Overall specific heat capacity for all-layers-together is termed here as 'effective specific heat capacity', which is calculated using Eq. (3).

1. Effect of Holding Mode Heat Leak

Considerations:

Case ID	Holding mode heat leak considered? (Y/N)	Initial ship-tank temperature (°C)	JBOG Compressor capacity (kg/hr)	Maximum allowed cooling-rate for ship-tank (°C per 20 minutes)
1A	N	-125	80,000	3
1B	Y	-125	80,000	3

17 AIChE Spring Meeting, San Antonio, TX

13

❖ This includes two cases, to compare LNG loading with holding-mode heat-leak (HMHL) considerations, and without HMHL considerations. Case-1A is without HMHL considerations, and Case-1B is with HMHL considerations.

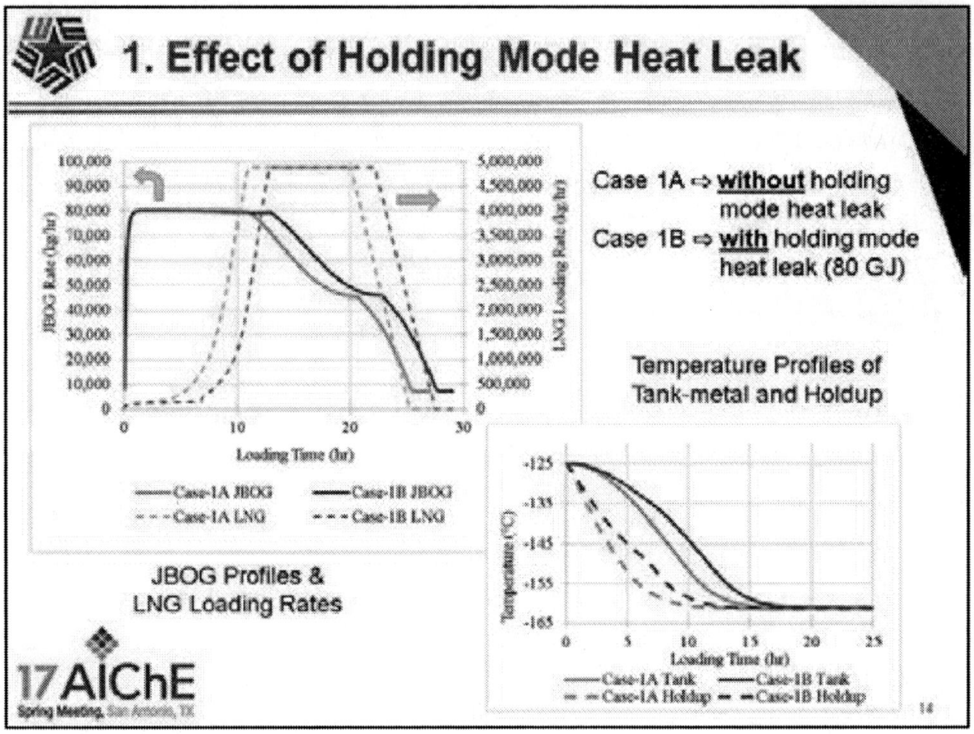

- ❖ Left plot shows LNG loading rate and JBOG rate with respect to loading time.
- ❖ Right plot shows cooling rate of ship-tank and LNG in the ship-tank

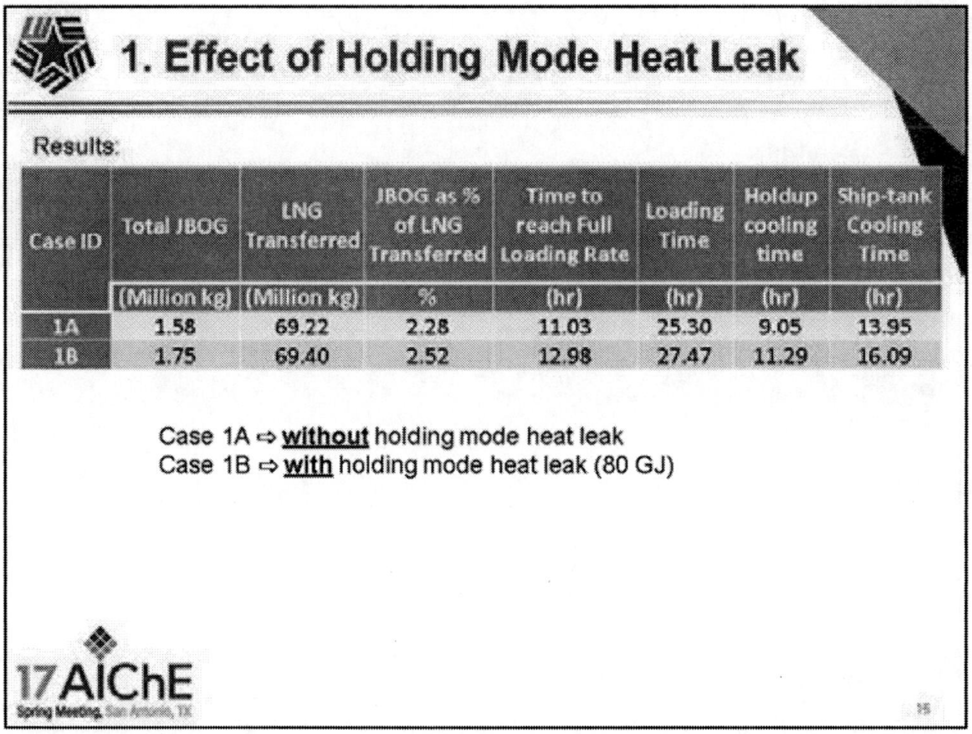

❖ The results for Case1A and 1B are shown here.

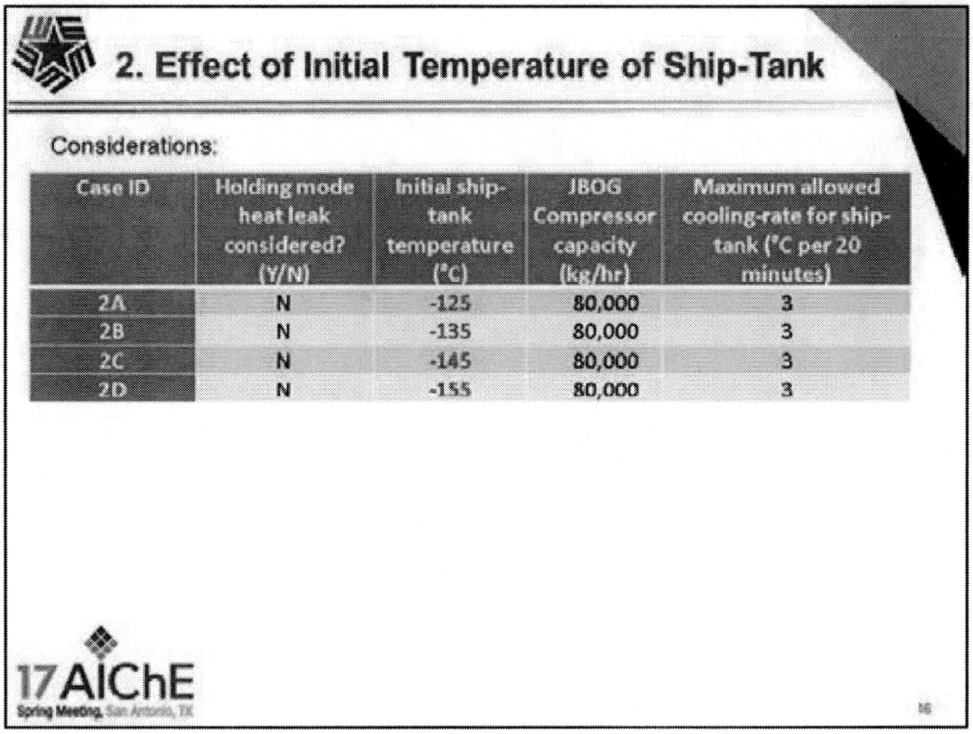

2. Effect of Initial Temperature of Ship-Tank

Considerations:

Case ID	Holding mode heat leak considered? (Y/N)	Initial ship-tank temperature (°C)	JBOG Compressor capacity (kg/hr)	Maximum allowed cooling-rate for ship-tank (°C per 20 minutes)
2A	N	-125	80,000	3
2B	N	-135	80,000	3
2C	N	-145	80,000	3
2D	N	-155	80,000	3

❖ This considers effect of initial ship-tank temperature. Case-2A, Case-2B, Case-2C, and Case-2D represent initial ship-tank temperature of -125, -135, -145, and -155 °C respectively.

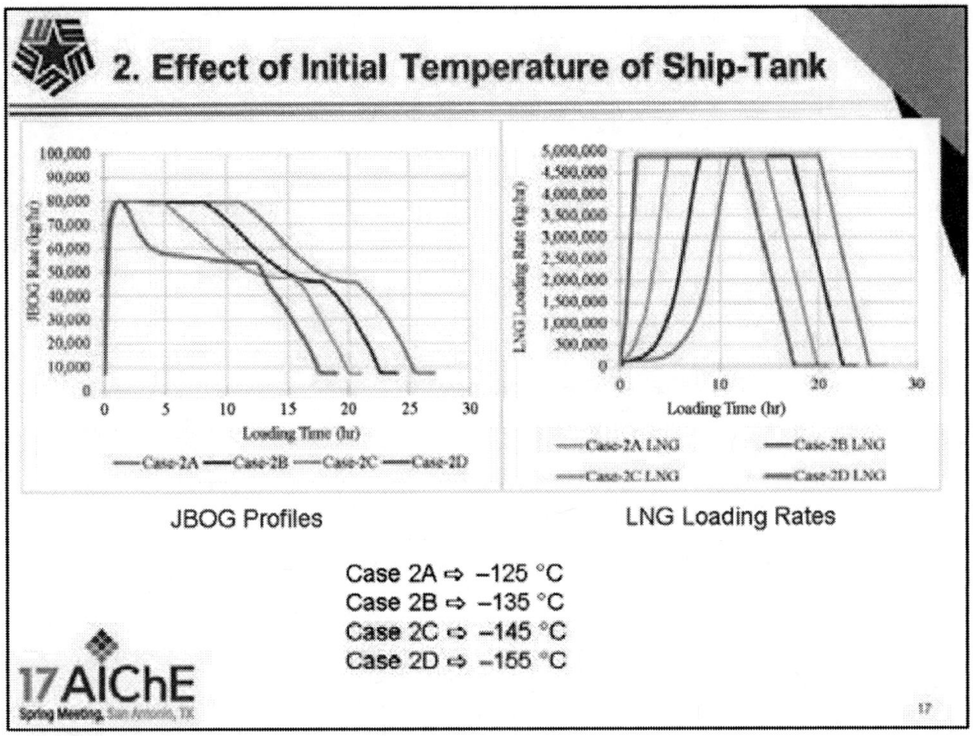

- ❖ Left plot shows JBOG profile, and right plot shows LNG loading rates for each case.
- ❖ Time required to achieve full LNG loading rate is about 11 hr, 8.1 hr, 4.9 hr, and 1.7 hr for Case-2A through Case-2D respectively

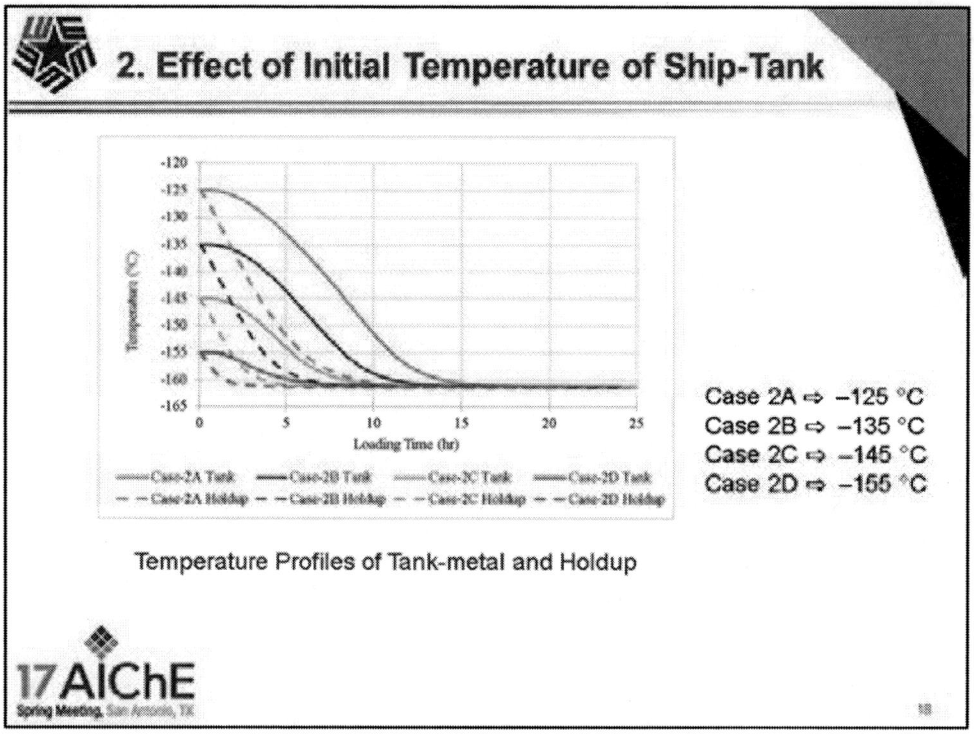

- ❖ This shows effect of ship-tank temperatures on time required to cool the tanks during LNG loading.
- ❖ For any particular case, tank cooling takes about 3.8 to 5 hours longer than cooling of tank-holdup to the temperature of loading LNG

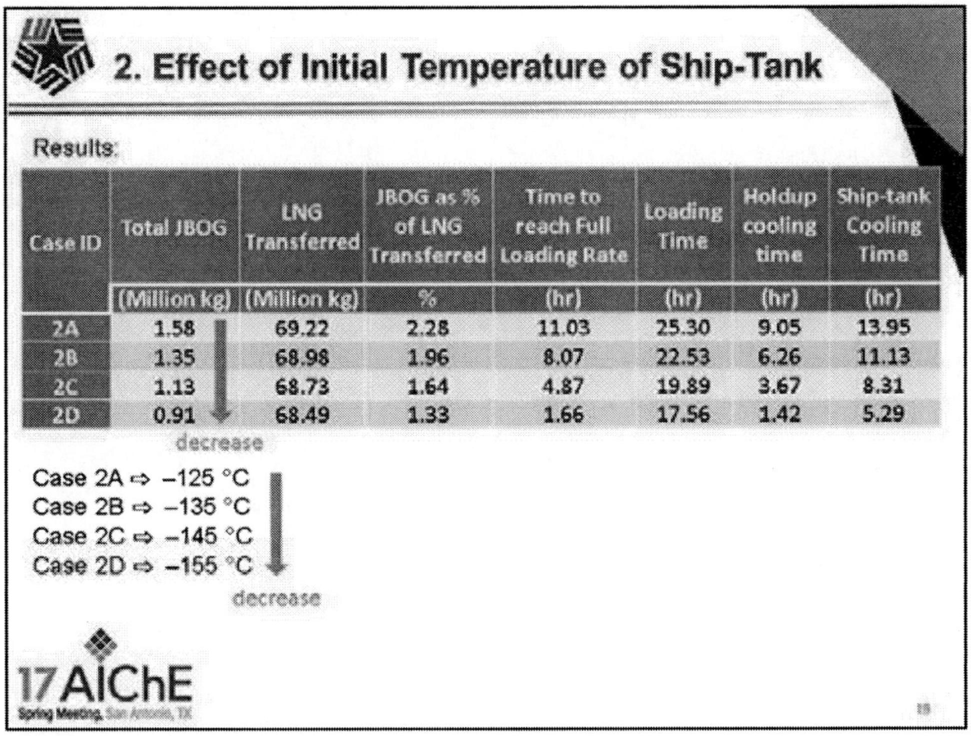

Case ID	Total JBOG	LNG Transferred	JBOG as % of LNG Transferred	Time to reach Full Loading Rate	Loading Time	Holdup cooling time	Ship-tank Cooling Time
	(Million kg)	(Million kg)	%	(hr)	(hr)	(hr)	(hr)
2A	1.58	69.22	2.28	11.03	25.30	9.05	13.95
2B	1.35	68.98	1.96	8.07	22.53	6.26	11.13
2C	1.13	68.73	1.64	4.87	19.89	3.67	8.31
2D	0.91	68.49	1.33	1.66	17.56	1.42	5.29

❖ Table shows the results for total loading time, total JBOG generated, total LNG transferred to ship, and cooling time for each case. Usually cooling is faster in the beginning and slows down later when temperature gradient decreases. To calculate cooling time for tank metal and holdup, the temperature of -160 °C is considered as criteria. Total amount of LNG transferred to ship for loading in Case-2D is only 1.1% less as compared to Case-2A; however, the corresponding total-JBOG decrease is more than 42%

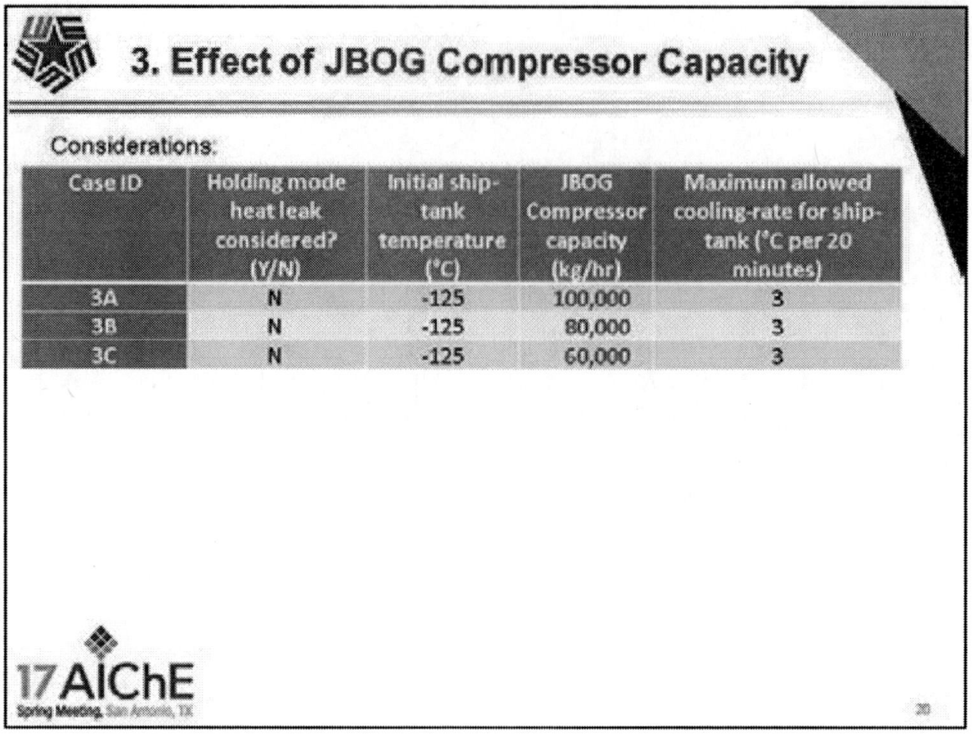

Case ID	Holding mode heat leak considered? (Y/N)	Initial ship-tank temperature (°C)	JBOG Compressor capacity (kg/hr)	Maximum allowed cooling-rate for ship-tank (°C per 20 minutes)
3A	N	-125	100,000	3
3B	N	-125	80,000	3
3C	N	-125	60,000	3

❖ This is to shows the effect of JBOG compressor capacity on LNG loading and JBOG generation. Case-3A, Case-3B, and Case-3C correspond to JBOG compressor capacity of 100,000 kg/hr, 80,000 kg/hr, and 60,000 kg/hr respectively

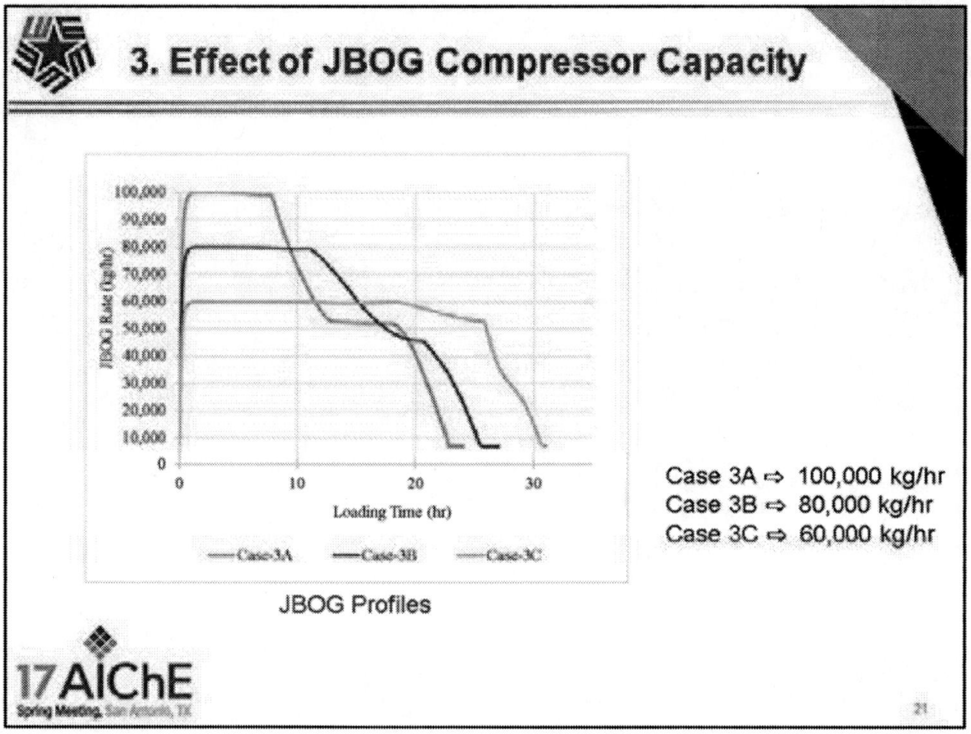

❖ For Case-3A, total-JBOG decrease is about 5.4% as compared to Case-3C, due to 7.7 hr decrease in loading time.

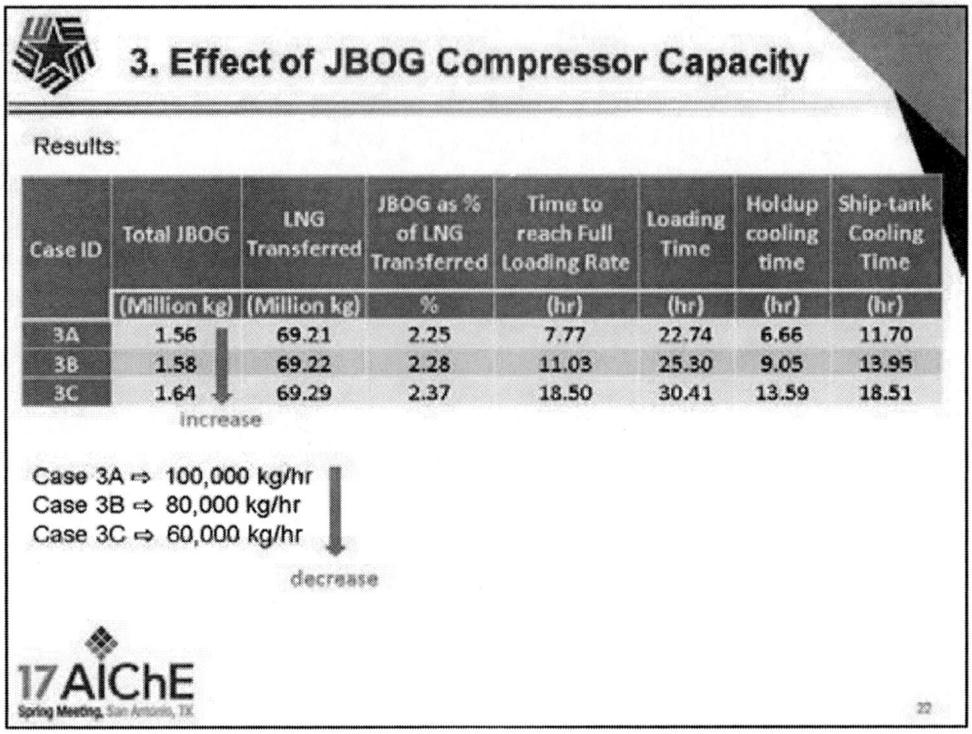

3. Effect of JBOG Compressor Capacity

Results:

Case ID	Total JBOG	LNG Transferred	JBOG as % of LNG Transferred	Time to reach Full Loading Rate	Loading Time	Holdup cooling time	Ship-tank Cooling Time
	(Million kg)	(Million kg)	%	(hr)	(hr)	(hr)	(hr)
3A	1.56	69.21	2.25	7.77	22.74	6.66	11.70
3B	1.58	69.22	2.28	11.03	25.30	9.05	13.95
3C	1.64	69.29	2.37	18.50	30.41	13.59	18.51

increase

Case 3A ⇨ 100,000 kg/hr
Case 3B ⇨ 80,000 kg/hr
Case 3C ⇨ 60,000 kg/hr

decrease

17 AIChE Spring Meeting, San Antonio, TX

22

❖ Higher capacity of compressors makes loading faster, and generate less JBOG. However, pipe size required to transfer JBOG will increase with the maximum JBOG flow rate. 24-inch pipeline is not sufficient to transfer JBOG at 100,000 kg/hr rate, when compressed up to 2.5 bar pressure. Higher pressure of JBOG and larger pipe size may permit higher JBOG transfer rates

4. Effect of Maximum Cooling-Rate

Considerations:

Case ID	Holding mode heat leak considered? (Y/N)	Initial ship-tank temperature (°C)	JBOG Compressor capacity (kg/hr)	Maximum allowed cooling-rate for ship-tank (°C per 20 minutes)
4A	N	-125	80,000	3
4B	N	-125	80,000	2
4C	N	-125	80,000	1.5
4D	N	-125	80,000	1

17 AIChE Spring Meeting, San Antonio, TX

23

❖ This is to study effect of maximum cooling-rate permitted for ship-tank. The restriction of the cooling-rate is to avoid thermal shocks to the tank materials. Case-4A considers the value of 3 °C cooling per 20 minutes. Case-4B considers the value of 2 °C cooling per 20 minutes. Additional values for the cooling rate are considered for the purpose of comparison. Case-4C considers value of 1.5, and Case-4D that of 1 °C cooling per 20 minutes.

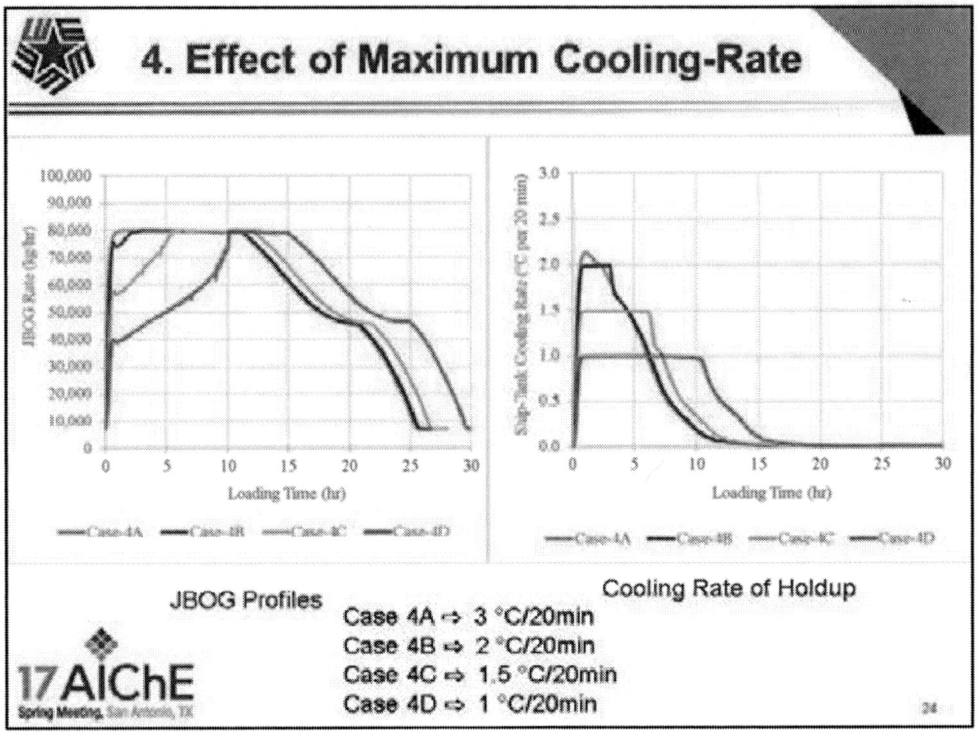

❖ It can be seen that JBOG rate did not reach compressor capacity limits approximately within first 2 hr for Case-4B, 5.5 hr for Case-4C, and 10 hr for Case-4D. Lower the value of maximum allowed cooling-rate, longer it takes to reach compressor limits, and also to complete the loading.

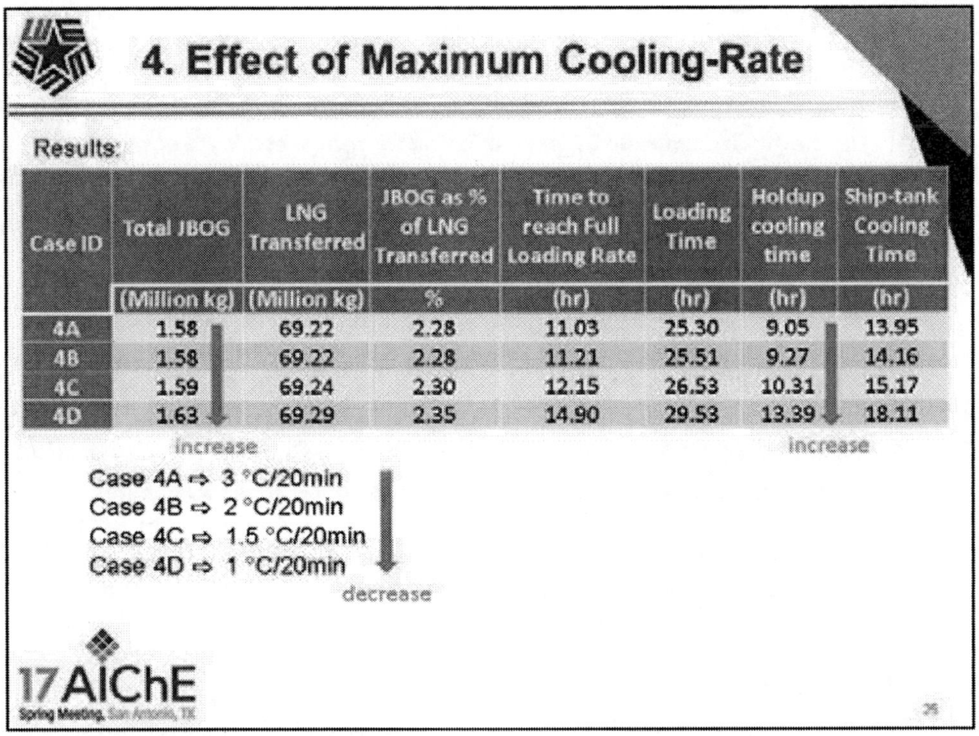

Case ID	Total JBOG	LNG Transferred	JBOG as % of LNG Transferred	Time to reach Full Loading Rate	Loading Time	Holdup cooling time	Ship-tank Cooling Time
	(Million kg)	(Million kg)	%	(hr)	(hr)	(hr)	(hr)
4A	1.58	69.22	2.28	11.03	25.30	9.05	13.95
4B	1.58	69.22	2.28	11.21	25.51	9.27	14.16
4C	1.59	69.24	2.30	12.15	26.53	10.31	15.17
4D	1.63	69.29	2.35	14.90	29.53	13.39	18.11

❖ Tank cooling takes about 3 hr longer in Case-4D as compared to Case-4A. The difference in loading time of Case-4A and Case-4B is only 15 minutes, because even in Case-4A (cooling limit of 3 °C per 20 minutes), the cooling rate reaches to value of only 2.1. It means, the compressor capacity constraint dominates the cooling rate constraint for Case-4A. For Case-4D cooling rate constraint dominates for the period up to first 10 hr of loading.

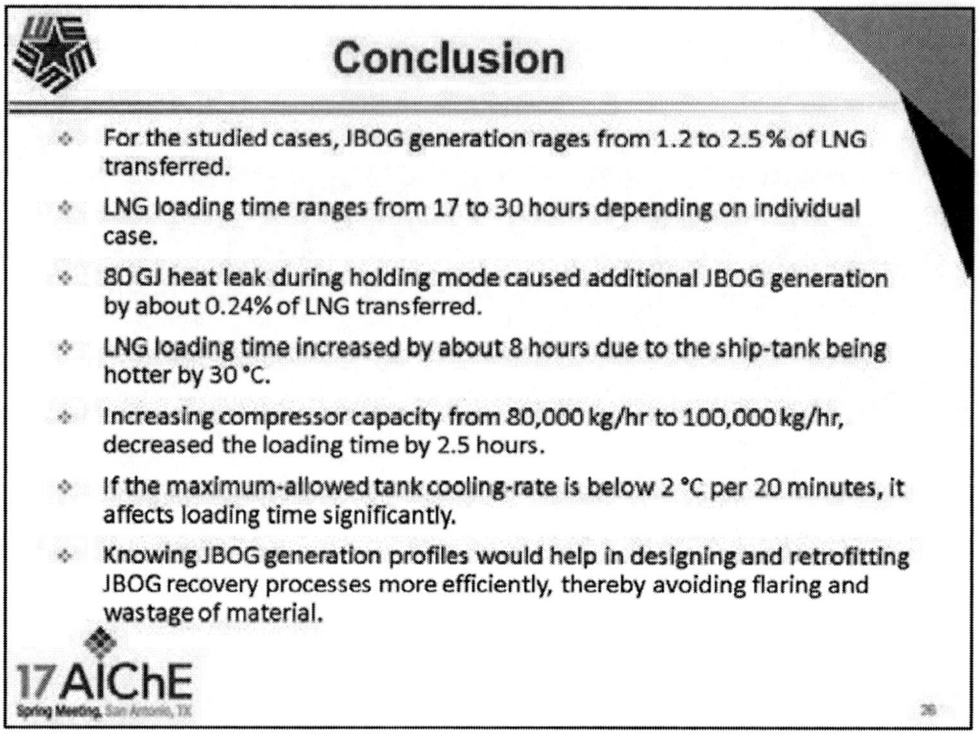

❖ Understanding JBOG generation would help in designing and retrofitting BOG recovery facilities in an efficient way. The study would help proper handling of BOG problems in terms of minimizing flaring at LNG exporting terminals, and thus reducing waste, saving energy, and protecting surrounding environments.

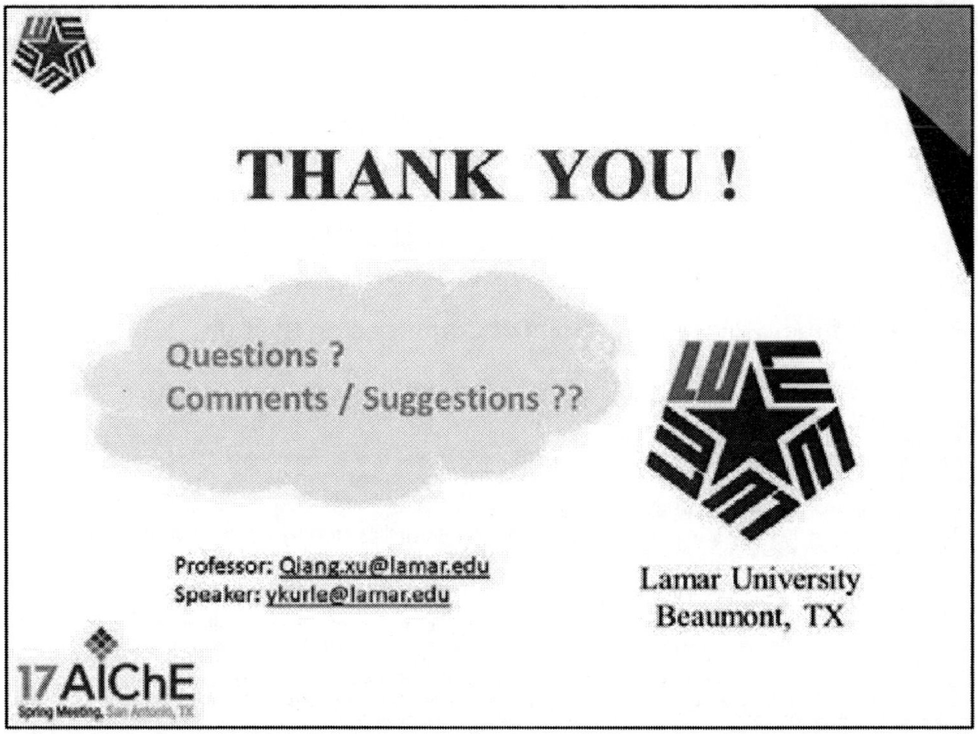

❖ Acknowledgement

This work was supported in part by Center for Advances in Port Management at Lamar University and Texas Air Research Center headquartered at Lamar University.

480196 NGL Product Storage and Boil Off Compression System

Abstract:

Natural Gas Liquid's (NGL) are components of natural gas that are separated from the gas stream in the form of liquids. NGL's typically refer to Ethane, Propane, Butanes and Pentanes. In a low gas price environment, NGL's are growing source of revenue for gas producers. Cryogenic distillation in a LNG facility is provided to separate methane and heavier hydrocarbons or NGL's from the gas stream. The heavier fraction is sent to the LPG fractionation unit where it is separated into C3, C4, and Condensate products.

In general LNG, C3 and C4 products are stored in double wall/full containment atmospheric storage tank. Facilities are provided for loading the LNG, C3, C4, and Condensate products to tanker ships. This paper focuses on the design, safety and controllability of C3/C4 product storage and associated Boil-Off-Gas generation in an NGL export facility.

The refrigeration and Boil-Off system converts warm and high pressure C3 and C4 liquid into cold and low pressure product streams for storage and transportation. A Direct or open system utilizes product as refrigerant, whereas an indirect or closed system utilizes a closed loop refrigeration system independent from product. Steady state simulation and design of open and closed refrigeration systems for C3 and C4 product storage are discussed in this paper, highlighting the advantages and disadvantages of the refrigeration system.

Dynamic simulation utilizing rigorous mathematical models has become an influential tool in the process design, design validation, relief system design, control system verification, startup support and troubleshooting. Many recent developments in the simulation software and technology have led to the development of large scale dynamic models. Dynamic simulation has been utilized for an existing LNG facility which has an open refrigeration system for cooling and storing C3 and C4 products. The Propane Boil-off-Gas compressor, which is an integral part of the open refrigeration system, has been modeled in detail.

Based on simulation results, modifications pertaining to safety and controllability of the C3 and C4 product storage facility are discussed.

OPERABILITY ANALYSIS FOR LNG COMPRESSOR STRINGS OPERATING IN PARALLEL

Laurence Casali, GE Oil & Gas, Florence, Italy
Lorenzo Gallinelli, GE Oil & Gas, Florence, Italy
Fabio Baldanzini, GE Oil & Gas, Florence, Italy
Marco Pelella, GE Oil & Gas, Florence, Italy

1 Abstract

Peculiar train configurations have been developed to satisfy new requirements in LNG industry in terms of availability of liquefaction plants. GE Oil & Gas has recently designed a new refrigerant train configuration composed by two identical strings each of them capable of 2x50% LNG production for PR-MR liquefaction process.

The peculiarity of this configuration requires dedicated studies to analyze critical transient scenarios such as trains start up, emergency shutdown, train trip in parallel configuration, malfunctions, etc. GE Oil & Gas performs dynamic simulations at two different levels.

The first level focuses on the emergency shutdown and start-up scenarios. The emergency shutdown analysis aims at evaluating additional protection (i.e. hot/cold gas bypass valves) for compressor surge/reverse rotation prevention. The start-up analysis is performed to evaluate driver capability. Further verifications are carried out to design and size process valves as well as validate check valves locations.

The second level addresses the entire compression train dynamic behavior by linking the compression system model to the actual compressor control system. This enhanced simulation environment, where also the interaction with liquefaction process is taken into account, allows to detect and solve in advance potential operability issues of parallel strings operation. In addition, start-up/shutdown procedures are optimized and the actual control software can be tested, validated and pre-tuned before commission activities.

The present paper covers dynamic simulation activities on the most critical transient scenarios for two PR-MR projects with different compression systems layout and different driver capability.

2 Train Mechanical Configuration

The GE compressor configuration described in the present paper is applied to precooled mixed refrigerant processes. The compressor train is composed by two strings, each of them capable of LNG production (2x50%), where PR and MR compressors are on the same shaft line.

Two projects are examined in the present work as examples of this configuration (see Figure 1). The heavy duty FR/7EA gas turbine is used as the main driver and a VSDS is used as helper motor: a 20 MW helper motor on Project 1 and a 24 MW on the latter. The three compressor casings are installed between the two drivers. The low pressure mixed refrigerant duty (LP MR) is accomplished by a horizontally split casing compressor MCL1405 on both projects. Then, project 1 employs a barrel back to back compressor 2BCL806 for medium and high pressure mixed refrigerant duties (MP/HP MR) whereas project 2 uses a barrel compressor BCL 806 for high pressure mixed refrigerant duty (HP MR). The propane duty is realized in a single horizontally split casing with 3 side-streams 3MCL1405 (4 sections corresponding to LP, MP, HP and HHP pressure levels) for project 1 and with 2 side-streams for project 2 (3MCL1404 with 3 sections corresponding to LP, MP and HP pressure levels).

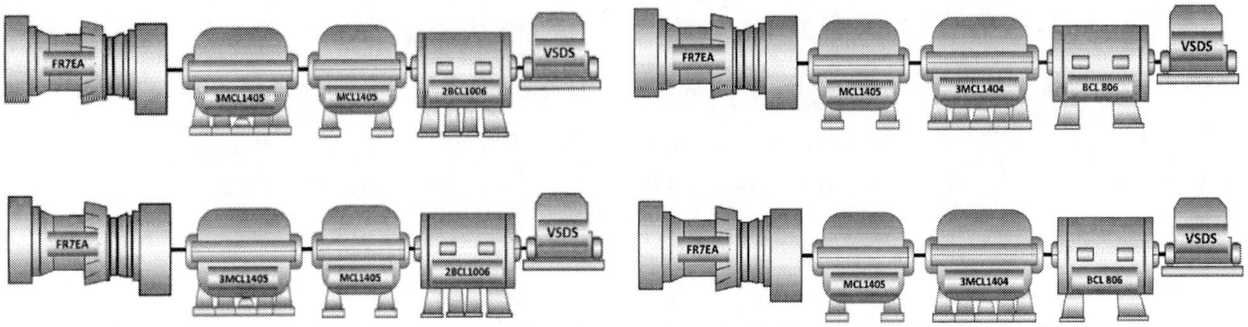

Figure 1. Train Configuration for Project 1 (left) and Project 2 (right).

This configuration will provide additional flexibility in plant management but requires special care in machinery design and operability.

3 Dynamic Simulation of Parallel Compressor Strings

3.1 Propane compressor system for Project 1 and 2

Propane (PR) compression system is shown in Figure 2, which is applicable to both projects apart the number of compressor sections as previously described. Each suction line is provided with suction isolation and throttling valves. On Project 2, suction check valves are installed downstream of each suction isolation valves. Each compressor section is protected by relevant anti-surge valves tie out from the common discharge header and tie in upstream of the relevant suction knock out drums.

Due to the parallel operation and to the limited speed range of the driver (from 95% to 101), each PR compressor is equipped with suction throttling valves manipulated for control scope (by the master performance/load sharing controller) or for handling critical sudden transient conditions (i.e. one string trip).

Propane system is also equipped with after-cooler and with quenching system on each compressor section. Quenching is mainly used during start-up scenario when one compressor string is isolated from the overall Refrigerant process and has to be put in service again. On Project 2, cross connection lines are installed to transfer propane from the isolated system to the operating one so as to speed up the full liquefaction production rate restoration. Differences between the Project 1 and 2 configurations are shown in red in Figure 2.

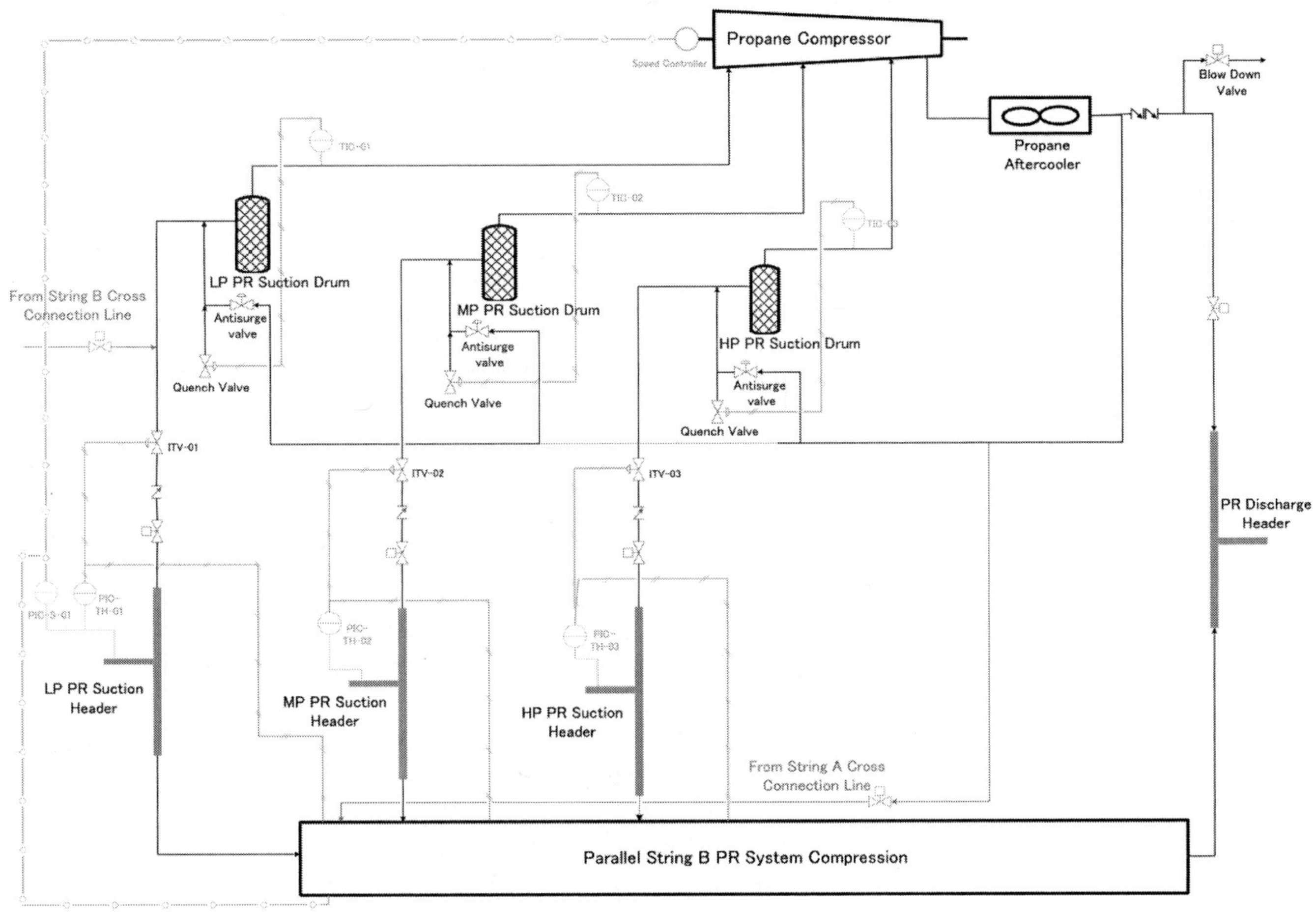

Figure 2. PR Compression System.

3.2 Mixed Refrigerant Compressor System for Projects 1 and 2

Mixed Refrigerant (MR) compression system is shown in Figure 3, which is applicable to both projects apart the number of compressor sections as previously described. The LP MR suction line is equipped with suction isolation valve and throttling valve. On Project 2, suction check valve is installed downstream of the suction isolation valve. Interstage valve is used to isolate each compressor section and a discharge isolation valve is located on the last discharge line.

Each compressor section is protected by the relevant anti-surge valve tie out from the relevant after-cooler outlet and tie in upstream of the relevant suction knock out drums.

Due to the parallel operation and to the limited speed range of the driver (from 95% to 101%), the LP MR is equipped with a suction throttling valve manipulated for control scope (by the master performance\load sharing controller) or for handling critical sudden transient conditions (i.e. one string trip).

Each compression section is equipped with after-cooler. On Project 1, the LP MR is equipped also with quench system used only during start-up scenario when one compressor string is isolated from the overall refrigerant process and has to be put in service again. On Project 2, cross connection lines are installed to transfer MR from the isolated system to the operating one so as to speed up the full liquefaction production rate restoration. Differences between the Project 1 and 2 configurations are shown in red in Figure 3.

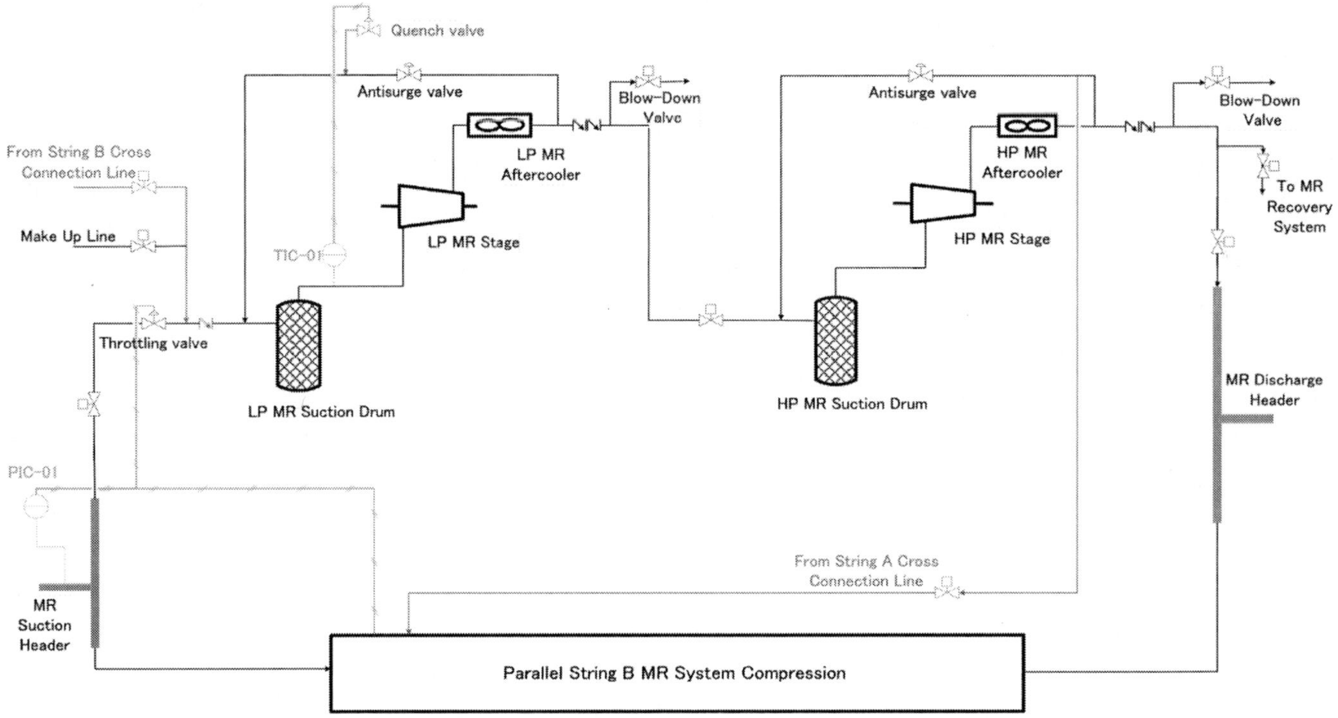

Figure 3. MR Compression System.

4 Operability Study

With the aim to optimize string parallel operation and operability a rigorous dynamic simulation study has been performed in two phases depending on whether the actual anti-surge/master performance/load sharing control system is linked to the process dynamic model.

4.1 Phase 1: Operability Analysis without the Actual Control System

Phase 1 of operability study aims to perform the following scenario:

- Emergency Shutdown (ESD) of one string, in order to verify the need of Hot/Cold Gas Bypass Valve (HGBV/CGBV) for compressor surge and counter-rotation prevention;
- Start up of one string, to check the driver capability.

Hysys® software has been used to develop detailed dynamic model of PR and MR compression systems. On Project 1, as previously described, suction check valves were not installed in both PR and MR systems. The consequence was that during the ESD transient, reverse flow to the main suction header was experienced as soon as the ASVs were opened. Based on the amount of flow recycled back to main suction header, compressor surge and risk of counter-rotation were affected negatively. The mitigation of compressor surge and the prevention of counter-rotation were achieved by installing HGBVs. Moreover, fast closing suction isolation valves were also installed to avoid oversizing the HGBVs and limiting the reverse flow to common suction headers (which may cause a sudden overloading of the companion string).

Similarly, on Project 2, ESD dynamic simulation showed that compressor Surge phenomenon did not comply with the acceptance criterion, making additional protections (HGBV) required on each compressor sections. The presence of suction check valves allowed avoiding any backflow to the common suction headers when Anti-surge valves opened. The risk of reverse rotation was also prevented; therefore, no fast closing suction isolation valves were needed.

Startup analysis showed that PR and MR compression system cannot be started from the settle out pressure, and the systems have to be partially depressurized in order to allow the FR/7EA and helper motor to accelerate the string to Minimum Operating Speed (MOS). The maximum starting pressure for PR and MR compressors shown in Table 1 were calculated by the dynamic simulation.

Start up scenarios have been instrumental in highlighting also sub-atmospheric conditions occurrence during the compressor acceleration ramp.

Table 1. Re-starting pressure for Propane and MR compression system.

Propane	2 - 2.6 bara
MR	2 - 2.6 bara

4.2 Phase 2: Operability analysis with the Actual Control System

In Phase 2, advanced dynamic model of the parallel Strings was developed and linked to the actual compressor control system (anti-surge/master performance/load sharing controller). This rigorous simulation environment allowed a deep study of transient conditions during several scenarios in order to mitigate potential operability issues, test control system, and pre-tune controllers. Phase 2 of operability study aims to perform the following scenarios:

- ESD of one string when both strings are running, in order to analyze the transient behavior of the companion string;
- Isolation valves failure of one string when both strings are running, in order to analyze the transient behavior of both strings.
- Turndown and normal shutdown of compression strings, in order to define and optimize the turndown and shutdown procedure.
- Start up of one string when the companion string is running, in order to define and optimize the start up procedure.

Thermodynamic time variant parameters in terms of pressure, flow rate and temperature were imposed at the battery limit of the dynamic model in order to have a more detailed interaction between the compression system and the overall refrigerant loop (evaporator/MCHE) dynamic behavior.

4.3 Advanced Dynamic Simulation on Project 1

This paper focuses on the ESD of one String and the isolation valves failure scenarios. During normal operation, strings operate in parallel sharing vapor coming from propane evaporators (PR compression system) and Main Cryogenic Heat Exchanger (MR compression system). As soon as the ESD of one string is initiated, the string is fully isolated from the parallel in few seconds due to the fast closing suction isolation valves (Phase 1 study). The complete isolation of the string causes a refrigeration system disturbance and a consequent transient condition where the vapor from propane evaporators and Main Cryogenic Heat Exchanger is routed to the running string only. The initial effect is an increase of both PR and MR suction headers pressure and flow rate with the risk to overload the still running string.

If the driver is not able to and/or is not fast enough (due to a limitation of electric grid capacity) to provide the required torque, the running string decelerates towards the minimum gas turbine speed and may trip due to the gas turbine under speed. If this condition is reached then a domino trip is experienced.

The dynamic simulation showed that domino trip can be prevented by partially closing the LP

MR suction throttling valve. If the string is running close to the MOS and the electric grid capacity is strongly limited, increasing also the gas turbine speed to Maximum Continuous Speed (MCS) may help to avoid gas turbine under speed trip. These changes must be maintained until the disturbance dampens out.

The isolation valve failure event is instead handled as described below.

Based on plant philosophy, the failure of any isolation valves required the complete isolation of the affected string. At the same time, the ASVs were opened (feedforward action) in order to protect compressors from surge and the gas turbine was decelerated slowly to MOS in order to reduce the fuel gas consumption. The complete isolation of the affected string leads to the same transient experienced during the ESD scenario. Hence, in order to prevent the ESD of the companion string, the same sequence as defined previously was implemented: partially close the LP MR suction throttling valve and increase the gas turbine speed to MCS. These changes must be maintained until the disturbance dampens out. Figures from 4 to 7 show the dynamic simulation results of MR suction isolation valve failure scenario on String B.

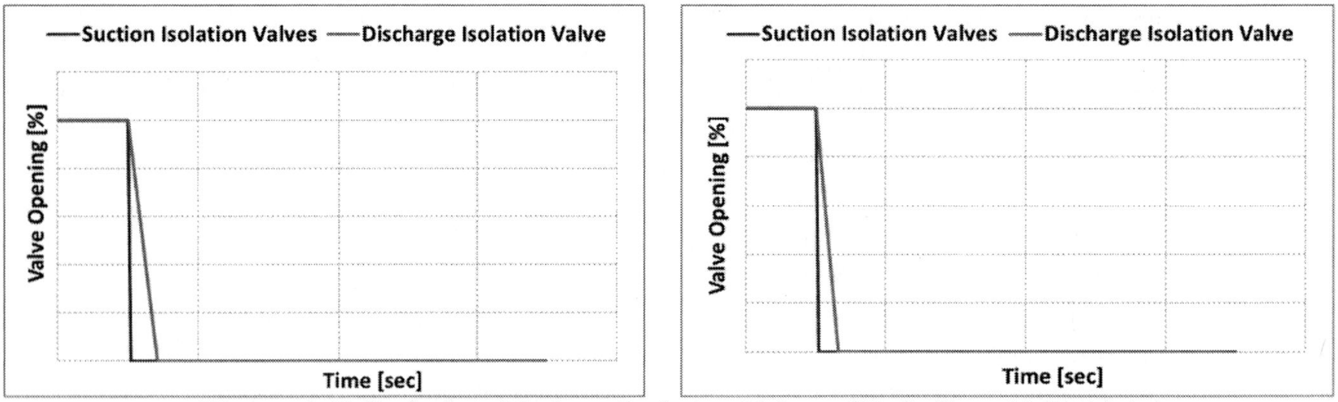

Figure 4. Fully Isolated String - Isolation Valves: PR (left) and MR (right).

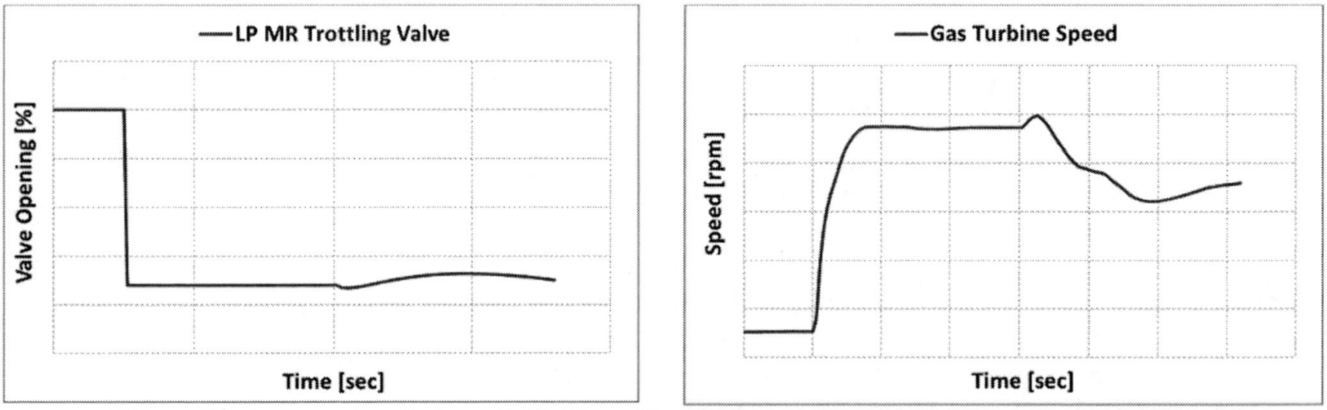

Figure 5. Companion String - MR Throttling Valve (left) and Gas Turbine Speed (right).

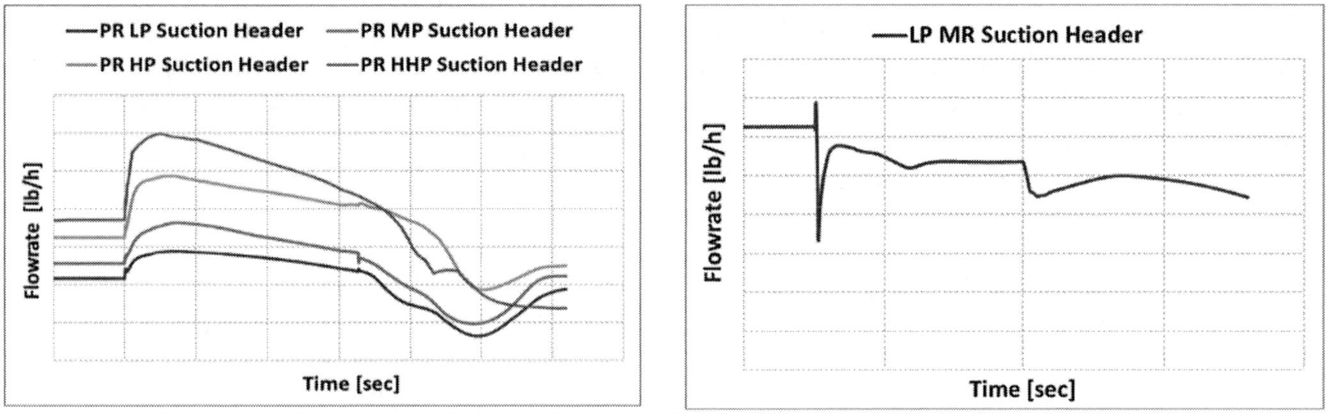

Figure 6. Companion String - Suction Header Flowrate: PR (left) and MR (right).

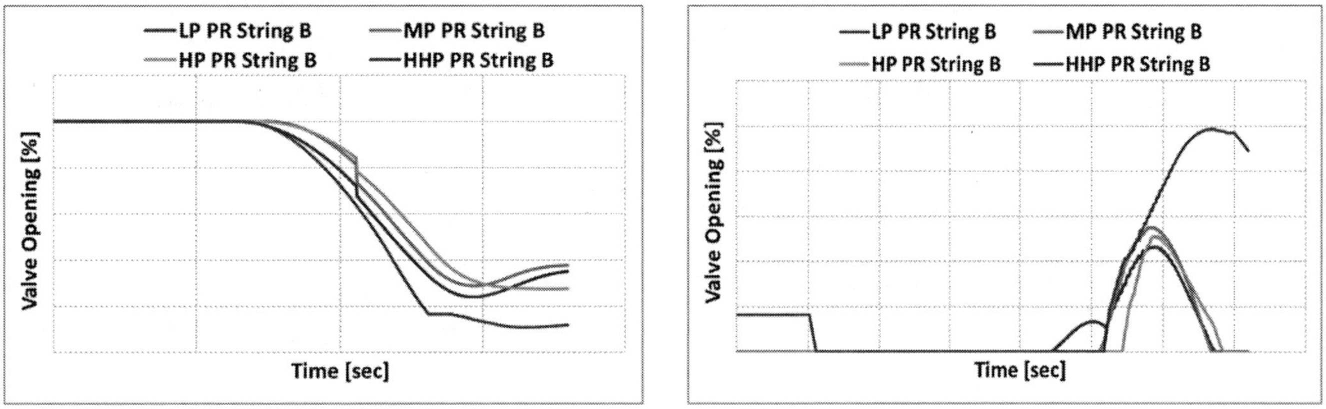

Figure 7. Companion String - PR Throttling Valve (left) and PR Anti-surge Valve (right).

4.4 Advanced Dynamic Simulation on Project 2

On project 2, the dynamic simulation showed similar results for the ESD scenario of one compressor string when both are running in parallel. The higher driver available power due to a bigger size helper motor allowed handling the sudden increment of absorbed power with higher margin versus the gas turbine underspeed trip threshold by stepping down the MR Inlet Throttling Valve. Also the helper motor power trend during the transient has been cross-checked with the plant electric grid capacity, without detecting any potential limitations.

Therefore, for project 2, it is covered only the isolated compressor string restarting scenario while the parallel string is online and the liquefaction process is operating at reduced production. Aim of the Start Up analysis is to investigate the restart of Compression String A (isolated from the Liquefaction Process) while Compression String B is running in parallel to cover the reduced capacity of the liquefaction process. It is crucial to guarantee the correct operability of the centrifugal compressors and an uninterrupted flow to the liquefaction unit. The main potential issue on the operating compressor string is the low flow transient which might enable the Anti-surge Controller till to make String B Anti-surge valve open and reach the full recycling condition. In this case, both compressor strings would be isolated from the liquefaction process. Even though the Main Cryogenic Heat Exchanger has a certain turndown capacity, this scenario must be avoided in order to restore the Liquefaction Process nominal conditions and minimize the duration of the transition to higher production rates.

With the aid of real Compressor Control System Emulator implemented within the dynamic model environment and consistent boundary conditions imposed to the dynamic model, a dedicated start up procedure has been developed and validated. The restarting sequence essentially outlines guidelines on how to handle anti-surge and inlet throttling valves, both on MR and PR compressors. Liquid Propane quenching has been also managed to speed up the propane compressor transient operation and avoid excessive liquid accumulation inside the Compression System.

The simulated String A restart scenario can be summarized through the following figures (from Figure 8 to Figure 11).

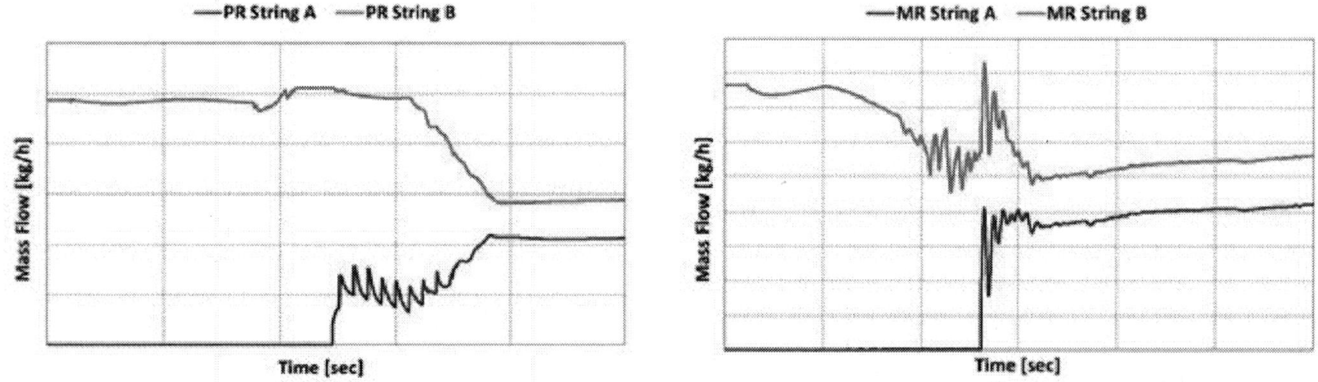

Figure 8. Flow rate through discharge check valves: PR (left) and MR (right).

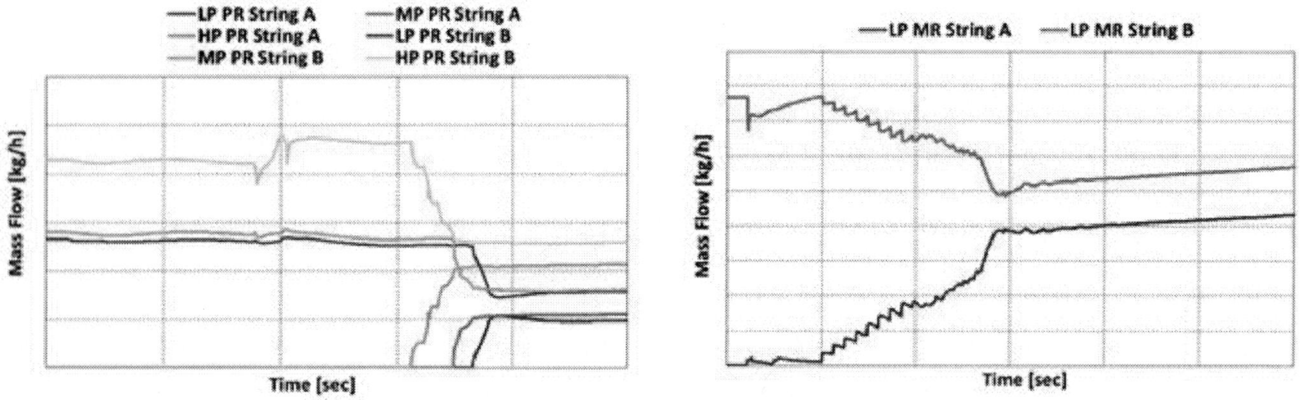

Figure 9. Flow rate through Inlet Throttling Valves: PR (left) and MR (right).

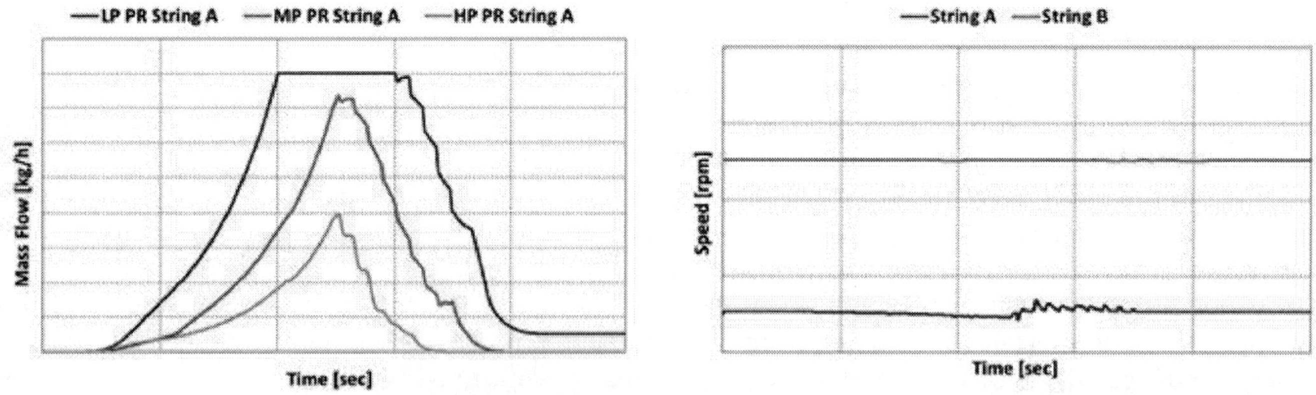

Figure 10. Flow rate through Quench Valve for PR (left) and String speed (right)

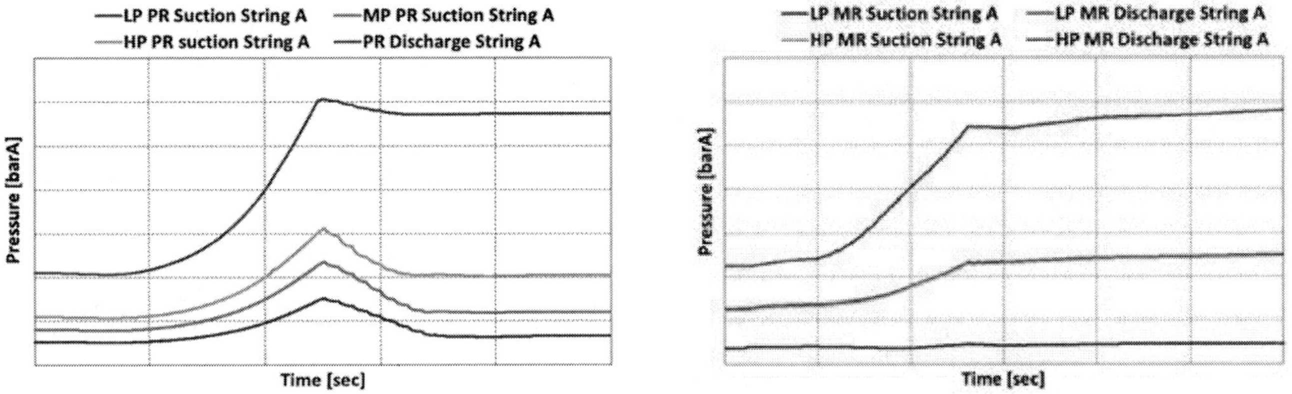

Figure 11. Compressor section pressures: PR (left) and MR (right).

When both compressor strings are back in service, the liquefaction process production can be increased till the desired rate: hence, the incoming flow rate to the compression system increases and the anti-surge controller starts to close accordingly each anti-surge valves up to fully closed position. Eventually, the master performance/load sharing controllers are activated and the pressure on the suction header is regulated by manipulating each inlet throttling valve.

5 Conclusions

Recent developments in LNG industry have driven manufacturers to a train configuration where each single string contains all main refrigerant compressors and therefore capable of LNG production, often replicated and operated in parallel to reach plant throughput targets.

Due to the peculiarity of this configuration, special focus is required on the compression system layout and compressor operability upon normal and transient operations.

As shown for two different projects, dynamic simulation is an instrumental and recognized tool to optimize string parallel operation and operability, validate and pre-tune the compressor control system and prevent potential upsets to the liquefaction process. In this paper a rigorous dynamic simulation study on two different projects has been introduced:

- Level 1: detailed model of the compression system aimed at mitigating the surge phenomenon, preventing the risk of counter-rotation and evaluating driver capability during start up.
- Level 2: advanced dynamic model of the parallel strings linked to actual control system to simulate scenarios like the ESD of one string when both are running, isolation valves failure and restart of the isolated string when the parallel one is in operation.

6 Acronyms

ASV = Anti Surge Valve
CGBV/HGBV = Cold/Hot Gas Bypass Valve
ESD = Emergency Shut-Down
HP = High Pressure
HHP = High High Pressure
LNG = Liquefied Natural Gas
LP = Low Pressure
MCHE = Main Cryogenic Heat Exchanger
MCS = Maximum Continuous Speed
MOS = Minimum Operating Speed
MP = Medium Pressure
MR = Mixed Refrigerant
PR = Propane
VSDS = Variable Speed Drive System

7 References

1 Pelagotti A., Baldassarre L., Meher-Homji C., Masani K. (2016), "Future Trends in LNG Turbomachinery", *LNG18*

2 Toci E., Pelagotti A., Gallinelli L., Casali L., Pelella M. (2016), "Machinery Design And Operability Study For LNG Single Liquefaction Line operating in Parallel", *LNG18*

AUTHOR INDEX

AIChE
120 Wall Street, FL 23
New York, NY 10005-4020

ISBN 978-1-5108-4130-7

CURRAN ASSOCIATES INC.
proceedings.com